EARTH
THEN AND NOW

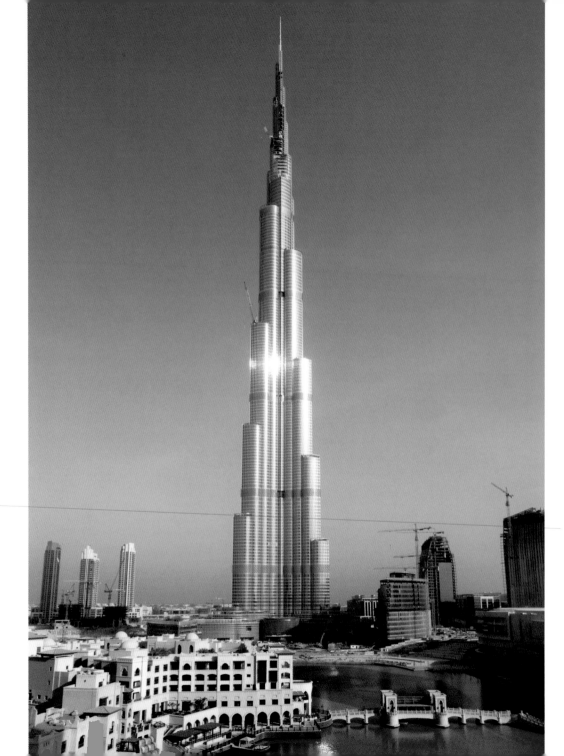

EARTH
THEN AND NOW
Amazing Images of Our Changing world

Fred Pearce

Foreword: James Lovelock

FIREFLY BOOKS

Burj Dubai The tallest building in the world, part of downtown
Dubai, was completed in early 2010.

Contents

A FIREFLY BOOK

Published by Firefly Books Ltd. 2010

Copyright ©2010 Octopus Publishing Group Ltd

This revised paperback edition published 2010

Publisher Cataloging-in-Publication Data (U.S.)

Pearce, Fred.

Earth then and now : amazing images of our changing world / Fred Pearce ; foreword, Zac Goldsmith.

[288] p. : col. photos. (chiefly col.),maps ; cm.

Includes index.

Summary: A collection of 250 stunning images showing locations around the world as they were and as they are now, with captions explaining the often breathtaking changes that have occurred in just a short amount of time.

ISBN-13: 978-1-55407-771-7 (pbk.)

ISBN-10: 1-55407-771-0 (pbk.)

1. Earth – Pictorial works. 2. Nature – Effect of human begins on – Pictorial works. 3. Human geography -- Pictorial works. 4. Global environmental change -- Pictorial works. I. Title.

550 dc22   TR655.P437   2010

Library and Archives Canada Cataloguing in Publication

Pearce, Fred

Earth then and now : amazing images of our changing world / Fred Pearce. -- Rev. pbk.

Includes index.

ISBN-13: 978-1-55407-771-7 (pbk.)

ISBN-10: 1-55407-771-0 (pbk.)

1. Earth--Pictorial works. 2. Nature--Effect of human beings on--Pictorial works. 3. Human geography--Pictorial works. 4. Global environmental change--Pictorial works. I. Title.

TR655.P43 2010   779'.361   C2010-902296-3

Published in the United States by
Firefly Books (U.S.) Inc.
P.O. Box 1338, Ellicott Station
Buffalo, New York 14205

Published in Canada by
Firefly Books Ltd.
66 Leek Crescent
Richmond Hill, Ontario L4B 1H1

Commissioning Editor **Jon Asbury**
Executive Art Editor **Yasia Williams-Leedham**
Senior Editor **Leanne Bryan**
Designer **Geoff Borin**
Editor **Andrew Puddifoot**
Picture Research **Giulia Hetherington**
Production **Peter Hunt**

Printed in China

Map Contents

Tent city Changing climate is pushing millions of the world's most vulnerable people to the brink. The El Hache refugee camp in Kenya's northeast province shelters 380 families, mostly nomadic herders who lost their livestock due to drought that extended across the Horn of Africa in late 2005.

Foreword

It is fashionable, at least in the Western world, to feel collective guilt: guilt about our ancestors having profited from slavery; guilt about racist or tribal feelings; guilt about the subservient place of women in society; guilt for feeling well fed and comfortable; most of all, guilt from winning. Fred Pearce's wonderful collection of then and now pictures adds a new dimension to our indulgence in guilt. We can now feel guilty for the partial destruction of the natural part of our planet.

I am sure that this is not what Fred intended and I hope that he shares with me the thought that humans are the Earth's most significant evolutionary child. We make mistakes, some of them horrendous, but how else can we learn? Natural selection picks from among the progeny those fittest to survive and that is why we are here, the latest result of Gaia's great experiment on the effect of intelligence on survival. The pictures of the site of the world's first atomic bomb test, and of the devastated Aral Sea, illustrate the havoc an intelligent animal species can cause; planetary-scale experiments are always risky but so far natural selection has saved the planet. Keep in mind the thought that one of the most dangerous experiments ever was the appearance of photosynthesizers as tiny cells floating in the ocean, perhaps as long ago as three billion years. These first algae made oxygen from water using the energy of sunlight to do it. A good and green thing to do you might think; unfortunately, at that time oxygen was about as poisonous as chlorine is now and as these algae flourished and spread there must have been a massive death toll and loss of biodiversity. In the course of time, life and the Earth adapted to oxygen and finally it became a benefit that allows large energetic animals like us to exist. The tiny photosynthesizers found a niche in the cells of plants and now provide the renewable energy that sustains all plant life. I like to think of humans with their intelligence as like those first algae: a life form that is very dangerous yet with near infinite promise; we are Gaia's gamble for a more secure old age.

As you look at these splendid pictures don't dwell on guilt, think instead of the good that we can do, not all the "nows" are worse than what once was. I prefer the vision of the poet Flecker who wrote one hundred years ago to a poet one thousand years hence: "...have you wine and music still and statues and a bright-eyed love, and foolish thoughts of good and ill and prayers to those who sit above?"

James Lovelock

April 2010

Introduction

Five years ago I stood on the promenade at Muynak, an old seaside resort on the shores of the Aral Sea in central Asia. Behind me was a fish-processing factory that had once sent canned fish across the Soviet Empire, from Warsaw to Vladivostok. No longer. Looking out to what should have been sea, I saw fishing boats abandoned on a beach that went on forever. The sea had disappeared 40 years ago, and its bed had turned into a new and entirely unexplored desert. Over the horizon, I was told, a remnant of what had once been the world's fourth-largest inland sea remained. But the shore was more than 100km (60 miles) away now and nobody in the town had ever gone to see it. I spotted a fox trotting where fish had once swum. In the far distance, a dust storm was brewing.

The Aral Sea died because Soviet engineers removed all the water from the two great rivers that once kept it full. They took the water to irrigate vast expanses of cotton farms – to grow the uniforms for the Red Army. The Russians have been gone for two decades now, of course, but the abstractions continue, and the fields are today growing cotton for shirts, jeans, and underwear on sale in almost every high street in the world. We are to blame now, as the sea continues to evaporate in the heat of the desert sun. It will probably be entirely gone within a decade.

UN scientists call the emptying of the Aral Sea the greatest environmental disaster of the 20th century. But I only really understood the scale of what had happened when I returned home from Muynak and looked at the pair of satellite images that appear in this book (*see* pages 54–5). They show a whole sea reduced to a toxic sump by human action. It is an unprecedented man-made change to the shape of the world. This book is full of such pairs of images, showing how our world has been changed. Most show bad things we have done, but by no means all. There are good stories here, too. This is the latest chapter in man's use and abuse of planet Earth. But all the images beg the question: what will happen next?

Humans have been making their mark on the landscape for a long time. Our hominid ancestors discovered how to make fire more than a million years ago, and they hunted animals across the plains of their first home, Africa. By the time of the last ice age our own species, *Homo sapiens*, had learned to combine these two skills to good effect, setting huge fires across the grasslands to drive animals toward their spears. It was the start of what has

Palm oil Clearing land for palm oil production is now a major cause of rainforest destruction, especially in southeast Asia. Palm oil is used as vegetable oil in products from chocolate bars and breakfast cereals to shampoo and as a bio-fuel. One Indonesian environment group fighting this loss, Yayasan Ekosistem Lestari, is encouraging local farmers to plant palm oil seedlings, like these, on fallow land cleared in past years.

A drop in the desert Wadi Rum is the largest wadi in Jordan. Lawrence of Arabia fought here, and the film of the same name was shot here. Engineers now pump water from beneath the desert, and pivot sprays irrigate circular patches of sand to grow grain. But the water is running out. Soon the desert will reclaim its own.

probably been mankind's biggest and most destructive impact on the land surface: deforestation. But we soon ceased to be hunters and gatherers alone. For at least 10,000 years, since the end of the last ice age, we have been farming. To make way for the crops needed to feed a fast-growing human population, we chopped and burned down ever-greater areas of forests, ploughed the great roaming grounds of wild animals, and drained huge areas of marshes. Nature was on the retreat.

But, thanks in part to the ephemeral nature of many human settlements, nature has demonstrated a remarkable ability to recover. Many apparently natural forests contain evidence that ancient human civilizations, of which we know virtually nothing, cleared them thousands of years ago. Afterward, nature returned. Modern explorers have rediscovered the

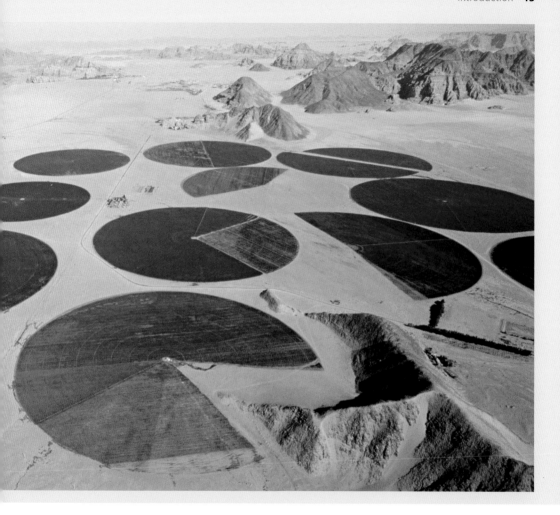

great Mayan ruins, for instance, buried in thick jungle in Central America. And those ruins are far from unique. Six hundred years ago there were cities with substantial populations in the Amazon jungle. Go back 1,500 years and the forests of central Africa were being turned into charcoal for metal smelting.

So, the good news is that nature can recover from the impact of human activity. But the bad news is that she has never experienced anything like the intensity of our current interventions. Today there are almost seven billion humans on the planet, a thousand times more than 10,000 years ago. And we now have advanced technology at our disposal. Once, we damaged small areas and then moved on; now very little of our planet is untouched by human occupation, and often the damage looks terminal.

Toxic cloud The Niger delta region of Nigeria was once rich in fish that provided protein for the millions of inhabitants of the coastal region. But the oil industry has moved in, polluting the waters and killing the fish. Now the locals are switching to meat. Their fishing nets are replaced by cattle pastures and foul abattoirs, such as this one in Port Harcourt, which is spewing toxic smoke across homes nearby.

Photographs cannot document those early footprints of humanity, but they do cover the past 150 or so years, during which our population and our impact has soared. In these pages you will see some of the consequences. In that time we have chopped down half the world's forests, doubled the area of land under the plough, all but eliminated large mammals, drained the majority of the world's marshes, and reduced most of the oceans' fish stocks and whale species by more than 90 percent. We have paved huge areas of the planet and broken much of the rest of the natural environment into tiny fragments by our extensive road-building. We have re-plumbed the rivers, plugging most of them with dams so that many no longer deliver water to the sea, and we have diverted their water instead through thousands of canals to irrigate fields. The demise of the Aral is only one – albeit the worst – outcome of that hydrological plunder.

We have dug deep into the bowels of the Earth. Look at the pictures of what we have done to Bingham Canyon in the USA and to Chile's Atacama desert – all in the pursuit of copper. And consider the way in which we have mined the South Pacific paradise of Nauru until there is, more or less literally, nothing left. Meanwhile, by releasing chemicals into the air we have burned a hole in the ozone layer and fundamentally altered nature's methods of recycling key elements such as sulphur, nitrogen, and – perhaps most dangerously for the future habitability of our planet – carbon.

Even seen from space, our handiwork is glaringly obvious. It is visible in huge reservoirs and megacities, in land drained from the sea, in whole coastlines that shift under our hidden hand, in dust storms that circle the globe, and in disintegrating ice sheets.

And yet there are things we can be hopeful about. Who can fail to be moved by the architectural magnificence of the Millau Viaduct in southern France, or by some of the dazzling modern cityscapes, or even, for its sheer *joie de vivre*, by the Paris beach? The construction of the Panama Canal killed thousands, but it remains an extraordinary monument to entrepreneurial and engineering endeavour. We should marvel, too, at the civil determination behind the rebuilding of San Francisco after the 1906 earthquake, the war-damaged Frauenkirche in Dresden, and the new Mostar Bridge in Bosnia.

Our creativity and inventiveness may have got us into our current environmental mess, but we must hope that they can get us out of it too. And there is evidence that they can. Look at some of the environmental rehabilitation projects undertaken after the closure of old mines. And think how the bleak, smog-filled industrial heartland of Bilbao has been transformed by an imaginative cultural vision. Scars can be healed.

Despite this, there is no denying that these chapters also contain evidence of tragedy, stupidity, venality, and short-term thinking in abundance. And, sadly, it is these misdemeanours that, for the time being at least, hold sway over the various examples of rehabilitation. Why, for example, care about coastal mangroves when they can be turned into toilet paper and the land annexed for prawn farms? Why worry about drying up the Aral Sea when there is cotton to be grown? Such human wilfulness has, of course, been immensely compounded by advancing technology. We have shovels that can lift 100 tonnes of ore in one scoop and dig holes up to 1km (3,300ft) deep. We plug rivers with concrete barriers hundreds of metres high, and drain marshes and defoliate rainforests because we can. Sometimes the scale of our endeavours is overwhelming. In Tokyo, for example, we have paved an area of more than 5,000sq km (2,000sq miles).

But we don't just inflict damage on nature. We do it to each other, out of fear and hate. The wars of the 20th century, I sincerely hope, will be looked back on by future generations as outbreaks of collective madness, never to be repeated. Thanks to the industrialization of warfare tens of millions of people have died, often for causes that were illusory or have long since ceased to matter. For me, trying to sum up in a couple of hundred words what exactly happened, and why, at Passchendaele – an event nearly a century ago – was almost unbearable. The devastation of Hiroshima and Nagasaki and the Siege of Leningrad also chill the soul. But perhaps the worst is over. Horrific as modern terrorist outrages such as the destruction of the Twin Towers are, they are orders of magnitude smaller in scale than the state-sponsored carnage of the two world wars. Nuclear weapons have not been used in anger since the summer of 1945. But, of course, while they continue to exist we remain under their shadow.

Nuclear weapons famously claimed to unleash the power of the sun. And, for all our inventiveness and technological sophistication, the forces of nature remain supremely powerful influences on the habitability of our planet. At times we have simply to stare in wonder: at the power of floods to wash away our world; at the force of earthquakes as the Earth's crust shifts; at tsunamis and avalanches that wipe out tens of thousands of people in their path; and at volcanoes that throw in our faces the molten contents of the Earth's core and shroud the planet in dust. One such volcano, 73,000 years ago, cooled the planet so much that our species came close to being wiped out.

And yet the truth is that we humans are beginning to change these great geological and planetary processes. Witness the extraordinary images of how the ozone hole opens up over Antarctica each southern spring. This is an entirely man-made phenomenon. The chemicals that make it happen did not exist in nature before we invented them in the early 20th century. Discovering the ozone hole in the mid-1980s was the moment when scientists realized that we truly could accidentally destroy our world.

Solar salvation Solar energy could be powering much of the world later this century. This solar power plant, located outside Seville in southern Spain, contains the world's first commercial solar "power tower". Six hundred movable mirrors direct the sun's rays to the top of the 115m (380ft) tower. The concentrated heat drives a steam turbine that generates electricity for 6,000 homes. European industrialists have plans to erect thousands of similar towers to harvest the sun of the African Sahara and power Europe.

And witness, too, the many images of melting glaciers and collapsing ice sheets. The Earth's stores of coal and oil and gas are the remains of carbon that nature captured from the air in much hotter times and gradually buried over tens of millions of years. That capture helped cool the Earth and keep it fit for the rise of humans. But now, by digging up and burning those fossil fuels, we are releasing that carbon back into the atmosphere. Every year, we unleash carbon that it took nature a million years to bury. Few people, beyond the perverse, can be truly surprised that our planet is warming fast as a result.

I am not among those who believe that we are doomed either by the pollution we have created or by our ransacking of the planet's resources. But nor am I indifferent to what we have done, as a significant minority of people still seems to be. The stakes are high. We do have the power to destroy our world. Nevertheless, I hope that you will see in these images some signs of our virtue as well as of our sin; of reasons for hope as well as for despair.

For me, one of the strangest pairs of images is that of Pripyat – a Soviet town built to house the workers at a power station in the Ukraine. The first of the two pictures looks like any other bleak landscape of tower blocks built by unthinking bureaucrats. The second image is, at first sight, similar, until you notice all the overgrown trees and weeds in the squares and streets. And the absence of people. It is like a post-apocalypse landscape.

And that, for the people of Pripyat, is what it is. For they worked at Chernobyl, the nuclear plant that caught fire in 1986. It released so much radioactivity that the town was evacuated and will not be safe enough to reoccupy for centuries hence. This is a human disaster, of course. But it has been a boon for wildlife. Down there among the trees are wild boar and deer, elks and wolves. They may be a bit radioactive, but they have revelled in the departure of humans and reoccupied the city. Nature has returned, just as it did after the fall of the ancient rainforest empires.

Is there a final lesson here? I think so. Nature is not as fragile as we think. She is resilient. With time, she may recover from the worst we can throw at her. It is we, ultimately, who are the fragile ones. Look at these pictures and fear not so much for nature: fear for us.

Allied bombing destroyed most of the centre of Cologne, Germany's fourth-largest city, in 1942. The twin spires of its famous Dom were among the few structures to survive. Behind, the Hohenzollern Bridge across the River Rhine lay badly damaged.

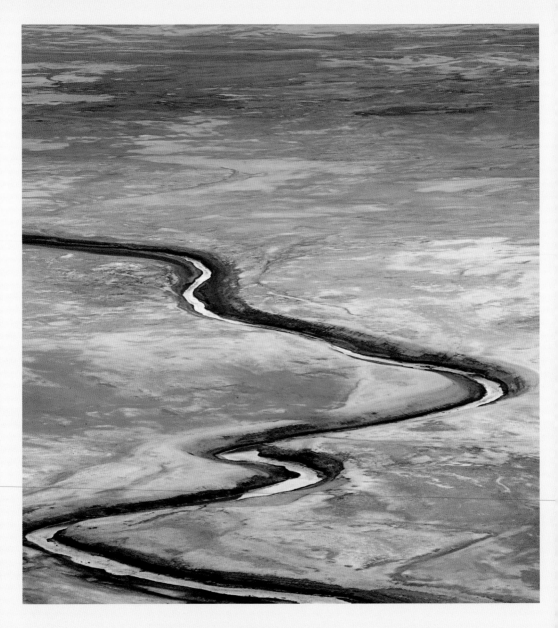

Bolivian drought The Milluni reservoir in the Bolivian highlands near La Paz has been emptied by a combination of drought and the disappearance of glaciers that once fed it. Reservoirs across the country, and in neighbouring countries like Venezuela, were at alarming low levels in late 2009 threatening both power and water supplies.

Environmental Change

Where does the wrath of nature end and the influence of mankind begin? As we watch deserts grow and ice caps melt, it is not always obvious. Nature is always on the move; never a passive spectator. But even so, this chapter reveals growing evidence that humans are implicated not just in obvious things, such as chopping down rainforests and draining wetlands, but also in apparently natural changes to the landscape.

Several of the examples involve rivers. Rivers flood naturally, of course. Nothing new in that. But many floods are as much to do with human activity as nature. In nature, rivers have floodplains over which they freely roam, but we have taken over much of this land, barricading it off from the rivers. No wonder, then, that great rivers such as the Elbe in Germany seek out the weaknesses in our flood defences and overwhelm them, causing floods far more destructive than anything that came before.

All rivers can suffer from low flows, too. But river abstractions, such as on the River Yangtze in China, can artificially empty rivers. And engineering on rivers such as the Rhine, intended to make it more navigable, has contrived to increase the risk of both low and high flows. Likewise, all rivers erode silt from mountains and carry it to the sea. But when humans are involved in deforestation of the hillsides, the silt flow increases. We see the consequences on the Yangtze in China.

All these changes are dramatic, but local. Mankind's newest and most pervasive impact is global climate change. Nobody now doubts that the world is warming. And few outside a hardened band of climate sceptics deny that the billions of tonnes of heat-trapping greenhouse gases that humans have put into the atmosphere are largely to blame.

In the past 30 years, for instance, climate change has doubled the area of the planet suffering from droughts. Combine that with bad environmental practice on the ground and you have a major crisis. In West Africa, reduced rainfall and water abstraction for irrigation projects has almost dried up Lake Chad. In northern China, drought and bad farming is eroding soils and creating massive dust storms that infest the capital and blow all the way to Canada.

The most unambiguous evidence of warming is the near-universal retreat of the world's glaciers. Among the examples we show here, the most famous is right on the equator, where Kilimanjaro, Africa's tallest mountain, is rapidly shedding ice. Within a decade it could be ice-free. The great polar ice caps, which have the potential to raise sea levels by tens of metres if they melt, are also under pressure from global warming. This chapter includes examples from Antarctica, and evidence that Arctic sea ice, too, is in full retreat and could be gone altogether in summer within a few decades. We are starting to see the consequences of all this in rising sea levels all around the world. From St Mark's Square in Venice to the paradise of the Maldives in the Indian Ocean, terra firma itself is under threat.

TRIFT GLACIER Switzerland

2002 The Trift Glacier is in the central Swiss Alps. A favourite with mountaineers, it carries ice formed on the 4,000m- (13,000ft-) high Wiessmies Mountain, and ends in a long tongue of ice that crosses a lake that was gouged out by the glacier during the last ice age. The tongue has been a permanent feature of the glacier since it was first scaled in 1855, and probably for much longer.

Retreating ice Only one year after the first picture was taken the ice tongue has been replaced by open water. The ice had retreated by more than 500m (1,640ft). A permanent path across it, used by generations of mountaineers, has disappeared. The melting was caused by exceptionally warm weather during the summer of 2003, which triggered a massive thaw right across the Swiss Alps. Glacier tongues such as this one proved particularly vulnerable in the face of a combination of warmer air and warmer water. Glaciologists say the Trift tongue is unlikely ever to re-form. After this picture was taken tour operators replaced the path with a suspension bridge across the lake. Meanwhile, filled with meltwater, the rising lake is close to breaking its banks, and engineers have been siphoning water in order to lower water levels and prevent a catastrophic tidal wave shooting down the valley.

GLACIER NATIONAL PARK USA

1912 The Glacier National Park in the US state of Montana is losing its glaciers. This image was taken two years after the park was created, when there were more than 100 glaciers. Blackfoot Glacier runs down the valley, and the Jackson Glacier comes in from the right, forming a tributary to its larger neighbour. Back then, the two glaciers covered some 8sq km (3sq miles).

Bare rock
By 2009, on the eve of the park's centenary, the scene was transformed. Both glaciers have retreated and they have become disconnected. The two glaciers today cover less than 2.5sq km (1sq mile) – not a third their former size. Meanwhile, in the foreground, ice has been replaced by bare rock and in places by vegetation, including the beginnings of a new forest on the ridge above the path of the former glacier. Nature is returning. The retreat of the ice is a sign of warming in the park. Today only around 50 glaciers remain. Three-quarters of the area once covered by ice and snow is now uncovered, or replaced by lakes of meltwater. Some glaciologists predict that Glacier National Park, one of the US's first formally protected areas, could have no glaciers at all by 2030.

MOUNT EVEREST Tibet

1968 The glaciers around Mount Everest in Tibet are melting. The top image, taken by Chinese scientists in 1968, shows the Middle Rongbuk Glacier, the area's largest. Tibet, the "roof of the world", contains the largest mass of ice outside the great polar ice sheets of Greenland and Antarctica. These glaciers are the sources for some of the world's largest rivers, including the Ganges in India, the Indus in Pakistan, the Yangtze in China, and the Mekong that waters southeast Asia.

Meltwater
As global warming diminishes the size of many of the glaciers, the flow of some of the rivers draining Tibetan glaciers is also in decline. That includes the Rongbuk, according to the chief lama of the Rongbuk Buddhist monastery, which is close to its source. By 2007 the toe of the Rongbuk Glacier had retreated by more than a mile. John Novis, the Greenpeace photographer who took the bottom image, was unable to reach the precise position where the 1968 picture was taken because landslides on the thawing mountain slopes have blocked the path. This picture is taken from less than a mile away.

RHÔNE GLACIER Switzerland

1932 The Rhône Glacier once filled the Grimsel Pass in the Swiss Alps, on the southern slopes of the Dammastock Peak. The glacier is a vital resource, providing meltwater that fills Lake Geneva 150km (93 miles) downstream, and forming the starting point for one of Europe's most vital navigation arteries and sources of hydroelectricity, the River Rhône. The glacier has been slowly retreating since the last ice age ended 10,000 years ago. Here, in the 1930s, it is less extensive than in the mid-19th century, when it reached right to the town of Gletsch in the valley bottom.

Unstoppable thaw

Since the 1930s the glacier has been disappearing ever more rapidly, largely as a result of the significant warming in the Alps. Particularly in the past decade, its retreat up the mountainside has accelerated sharply. Local towns depend on the glacier to attract visitors, and each summer, now, the people of Gletsch rush out to cover the glacier with huge insulating blankets to try to protect its tongue from further melting. But to little avail. The retreat goes on. In the short term the thaw has increased river flow, but in the long run, as the glacier comes close to disappearing altogether, the regular supply of water from it will disappear. Then Lake Geneva and the River Rhône will be entirely dependent on rainfall.

UPSALA GLACIER Patagonia

1928 Eight decades ago the glaciers of Patagonia in southern Argentina were mostly advancing. Among them was South America's largest, the Upsala Glacier. It was named after the Swedish city whose university carried out the first surveys of the glacier, although this picture was taken in 1928 by one of the earliest Argentinian explorers of the Patagonian glaciers, Mar'a Albert of Agostini, "the Father of Patagonia". The glacier's river of ice flowed down from the mountains into the Argentino Lake, which it blocked and filled with its icebergs.

Open water The glacier has retreated. It no longer blocks the Argentino Lake, and what was once ice has turned to water. Upsala Glacier is now almost 4km (2.5 miles) shorter than in 1928. The retreat has been unrelenting since 1978; the glacier lost almost 0.8km (0.5 mile) in 1994 alone. Global warming has almost certainly played a part in this, but the change may have been accelerated because less ice is flowing down the glacier today. Nonetheless, all told, the Patagonia glaciers and ice fields have lost 2,000cu m (70,600cu ft) of ice annually in recent years – a faster rate of loss than anywhere else on Earth.

STIEREGG RESTAURANT Switzerland

2004 The Stieregg restaurant had been in business since 1952, more than 1,600m (5,200ft) up in the Swiss Alps. It was a regular destination for people heading toward the Grindelwald Glacier on the slopes of the Eiger. But the glacier has been in retreat for some years. And, more critically still, the permafrost beneath it is melting. Many mountainsides like this are held together by the permanently frozen soil. But as temperatures rise the soil melts and what was once as solid as rock turns into an unstable scree slope. It doesn't take much to turn that slope into an avalanche as lethal as any made of snow.

Mountain erosion Rising temperatures have sealed the fate of the Stieregg restaurant. First the thawing mountainside above it began to unleash boulders down the slope. Millions of tonnes of rock have broken off the side of the Eiger in recent years. Then, during the height of the summer melting season in 2005, the swollen river in the valley beneath the restaurant began to erode its banks. At the end of May, some 300,000cu m (10,600,000cu ft) of thawed soil crashed into the valley and the restaurant was left hanging over the edge of a rapidly eroding precipice. The local authorities eventually set fire to the building to prevent it from adding to the debris below – another victim of global warming.

MOUNT KILIMANJARO Tanzania

1974 There has been ice on top of Mount Kilimanjaro for at least 12,000 years. Although the mountain is only 350km (220 miles) from the equator, the air at its 6,000m (19,000ft) summit has been cold enough to turn the rain to snow and keep it frozen all year. The snow and Kilimanjaro's status as the highest peak in the world accessible to ordinary tourists have made the mountain Tanzania's top tourism moneyspinner. But Kilimanjaro, which means "shining mountain", is losing its sparkle. The ice fields in its volcanic crater and the glaciers down its flanks have been disappearing since they were first measured in 1912. When this picture was taken in 1974, the ice was already in retreat.

Melting ice cap Today the ice is disappearing faster than ever. Early ice loss was probably down to declining snowfall caused by deforestation on the slopes. But today, warmer temperatures are melting the slopes fast. More than 80 percent of the ice is now gone, and on current trends the rest will melt away by 2015. Around the world, tropical glaciers are disappearing – from nearby Mount Kenya to the Peruvian Andes and the peaks of New Guinea. They say you can climb Kilimanjaro in trainers now. But, paradoxically, the mountain is becoming more dangerous, because the disappearance of ice and melting of permafrost are loosening rocks and causing more and more landslides.

LARSEN B ICE SHELF Antarctica

January 2002 The Larsen B ice shelf was an area of floating ice roughly the size of Luxembourg and connected to a peninsula of Antarctica that juts out into the South Atlantic. Glaciers in the mountains visible on the left of this image fed ice onto the shelf. Small icebergs would periodically break off from its outer edge, but the shelf itself had existed for at least 10,000 years, and looked secure enough when this picture was taken on 31 January 2002.

Shattered Appearances can be deceptive. The Antarctic peninsula, and the waters around it, had been getting warmer. The shelf had been melting from beneath, and pools of liquid water had formed on its surface. In mid-February 2002 – the warmest time of year in the southern hemisphere – the ice abruptly shattered. By 23 February a series of huge icebergs could be seen breaking away, and the shelf coastline had retreated.

New shoreline By 7 March most of the floating ice shelf had disintegrated. Numerous huge icebergs had begun heading out to sea, and the light-blue area was a large new bay full of the shattered remains of the ice shelf, gradually melting. In all, 2,600sq km (1,000sq miles) of ice shelf disappeared.

ARCTIC OCEAN

2005 The ice floating on the Arctic Ocean extends in winter and retreats in summer, reaching its minimum in early September. But the amount of ice that survives the summer has been diminishing ever since the first satellite images recorded the changes in 1979. Warming of both air and oceans has been taking its toll. By 2005, when this photograph was taken, summer ice had gone from many of the passages and gulfs of the Canadian Arctic and from ocean north of eastern Siberia. The total area of permanent sea ice had diminished by more than a fifth, with 2005 showing the smallest ever recorded ice cover.

Tipping point

But that was just the start. By September 2007, an even more dramatic change had occurred. Ice had disappeared from a much larger area north of Siberia and Alaska. There was open ocean almost to the North Pole. Probably for the first time in thousands of years, ships could sail both through the Northwest Passage beyond mainland Canada, and the Northeast Passage north of Siberia. Some believed that the Arctic Ocean had reached a "tipping point" and could soon be ice-free in summer. Others said the effect of global warming on the ice had been exaggerated by local warming effects caused by winds and ocean circulation that might not be repeated. Either way, it is a very long time since the Arctic ever looked as blue as this.

QINGDAO China

2008 Qingdao harbour in Shandong province has been a victim of Chinese industrial pollution. It opens onto the Yellow Sea between China and Korea, one of the most polluted seas in the world, known for its periodic "dead zones". Gorged on organic waste from sewage and industry, the harbour has often filled with green algae. In July 2008, the rotting algae choked the coastline for miles. The harbour was earmarked to be the main waters for sailing events at the 2008 Olympics in a few weeks' time, so the authorities decided on emergency measures. Here the mounds of algae are loaded into trucks. As well as heavy machinery, more than 10,000 local people lent a hand to clean up the algae in preparation for the Games.

Clean-up A year later, in July 2009, and the improvement has proved to be permanent. On the same stretch of harbour, the rotting algae has been replaced by a sandy beach, where bathers gathered one Sunday. The environmental improvements begun for the Olympics, which including moving a shipyard and shutting other sources of pollution, have been a boon to the city. At the end of 2008, Qingdao was voted among the seven best places to live in China.

LAKE DONGTING China

May 1998 Lake Dongting is a permanent feature on the floodplain of the River Yangtze in central China. The lake (seen in blue in this infra-red image) covers 2,600sq km (1,000sq miles), an area larger than that covered by Greater London, England. But it is less than half the size that it was a century ago. Since then, farmers have drained huge areas of former lake bed for rice paddy and much of the rest of the lake has filled with silt brought down by the Yangtze from deforested and eroding hillsides upstream in remote areas of western China. Most of the silt is covered in vegetation and shows up in the picture as red.

Sudden flood Each summer the monsoon rains swell the Yangtze, and farmers around Lake Dongting face the risk of inundation. But in 1998 the flooding was catastrophic. Between the first picture, taken in May, and this second in August, the lake has grown to several times its normal size, engulfing rice paddy and other low-lying areas. The flood killed 3,000 people and destroyed five million homes. Its severity was partly due to exceptional rains, but partly also to the denuded hill upstream that can hold less rainwater today than in the past. After the 1998 floods, in an attempt to prevent a repetition, China banned all logging of natural upland forests.

VENICE Italy

2001 St Mark's Square attracts the biggest crowds in one of the world's most popular tourist cities, Venice. Millions flock here to see the ancient basilica and sit at café tables in the piazza. Venice is unique. The former capital of a hugely successful merchant empire, it is constructed on wooden piles in a lagoon at the head of the Adriatic Sea. There are no roads or cars here, only boats travelling the ubiquitous canals that are its thoroughfares. But the water all around the city is also its biggest threat. And St Mark's Square, located right by the Grand Canal, is the city's lowest point.

High water

Venice has always been at risk of flooding, but man-made changes to the lagoon, and rising global sea levels, have greatly increased the danger. Once, the lagoon was rarely more than a metre deep, and its mud banks absorbed the force of the waves, keeping out high tides. Today waves surge into the lagoon along shipping channels dredged for cruise ships and other vessels. Many of the mud banks have been eroded. In the past three decades the average high tide in the lagoon has risen by almost 50cm (20in), flooding basements and washing across St Mark's Square around 50 times a year.

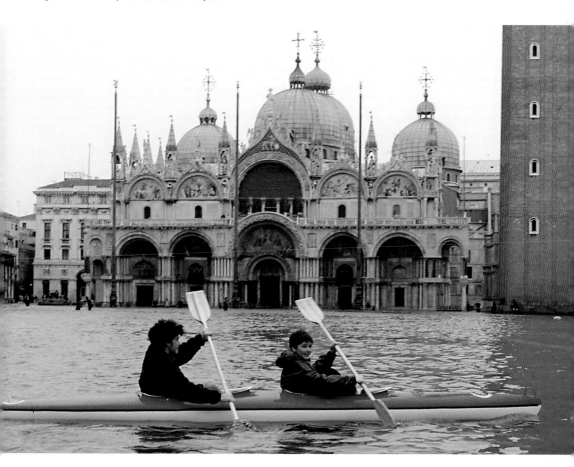

HITZACKER Germany

1999 The medieval German town Hitzacker sits on the River
Elbe where it joins the River Jeetzel in Lower Saxony. It is a beautiful
setting. This small settlement, with its many half-timbered buildings,
is surrounded by extensive flood meadows that are a paradise for
birdlife. Many of its 5,000 inhabitants can have boats moored close to
their houses. Around town, however, are signs of a more troublesome
side of this fluvial idyll. Marks everywhere reveal that high water often
enters the town, flooding its cellars and alleyways.

Frequent flooding
Inundation of this sort is happening more and more at Hitzacker. In August 2002 CNN came to the town to film exceptional floods rushing down the Elbe, but this picture was taken four years later in April 2006, when Hitzacker was again swamped after heavy rains combined with snowmelt in the Krkonose Mountains of the Czech Republic to swell the river. The chancellor, Angela Merkel, toured the town as the Elbe hit the highest marks ever recorded. Environmentalists blamed dykes and developments upstream that cut the river off from its floodplain and pushed river levels ever higher. Vulnerable towns such as Hitzacker paid the price.

LAKE CHAD Africa

1972 Lake Chad is a giant evaporation pond that drains the southern fringes of the Sahara desert. The lake is extremely shallow, averaging only 2m (7ft) or so deep. So its size changes dramatically as rainfall fluctuates in its catchment area. It was for a time the world's sixth-largest inland sea and straddled parts of four countries: Chad, Cameroon, Nigeria, and Niger.

Drying out
But in recent decades the rains in this region have been fitful. River flows into the lake have declined. Meanwhile, big irrigation projects have taken water from the rivers, reducing their flow further. Lake Chad has been drying out and its shoreline has retreated from Nigeria and Niger altogether. Traditional farmers around the lake have long since learned to cope with this constant change. They simply move with the lake edge. But large, engineered irrigation systems have suffered. The British-designed South Chad Irrigation Project in Nigeria, which was built in the 1960s to take water from the lake, is now high and dry, with its lowest intake tens of kilometres from the nearest shore.

POTOSI Venezuela

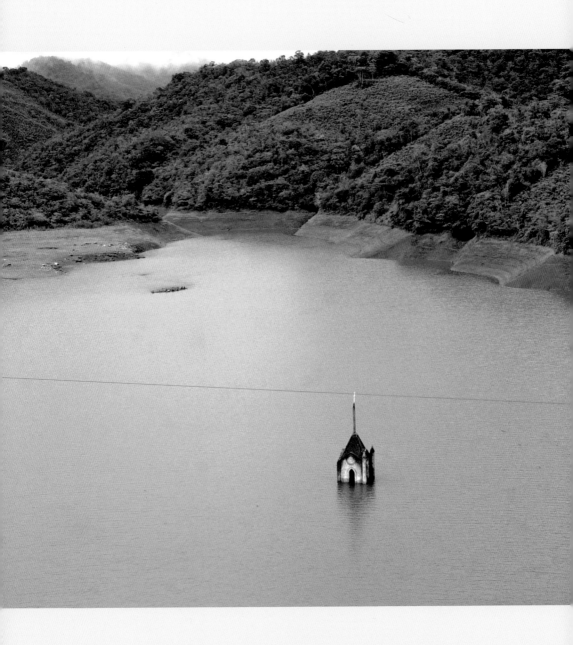

2009 Potosi town in the Venezuelan state of Tachira in the Andes disappeared in 1985, when a hydroelectric dam flooded the Uribante valley. Its 1,200 residents were forced to move. From time to time, as the turbines ran and water levels in the Uribante-Caparo reservoir fell, the peak of the 25m (80ft) church tower emerged from the waters – a tantalizing glimpse of the old town.

Drought A year after the first picture, in February 2010, a drought had hit the Venezuelan Andes. More and more of the ruined church was exposed, until even the floor of the nave was dry. Former residents returned and visited graves in the churchyard. But this evocation of the past has been a disaster for the country's power production. When this picture was taken the hydro-plant was only generating at 7 percent of its capacity. Venezuela, which normally gets two-thirds of its power from hydro-dams, faced an electricity crisis – an ominous sign of how vulnerable some countries are to fluctuating climate.

RIVER RHINE Germany

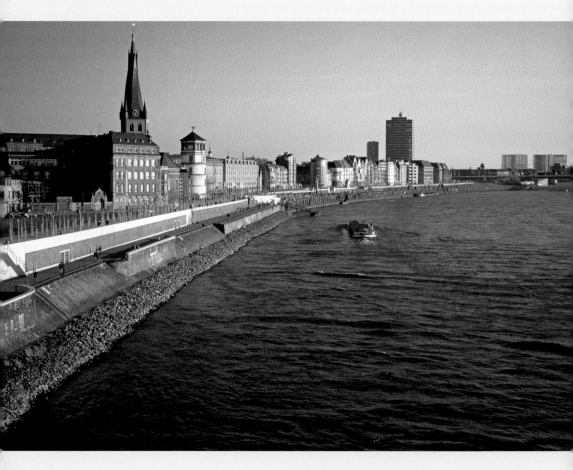

1990s Dusseldorf is one of the great German industrial cities. Once a small market town on the banks of the River Rhine, it has grown into a large and prosperous metropolis since the river was re-engineered in the 19th century as a single waterway navigable all the way from the North Sea to the Swiss border. In good times the modern engineering works well. Giant barges and even ocean-going ships make their way swiftly up and downstream on their way to the Ruhr industrial region, for which Dusseldorf is the financial centre. In this image the river is full, providing a fine backdrop for the old town, with the city's television tower overseeing everything in the background.

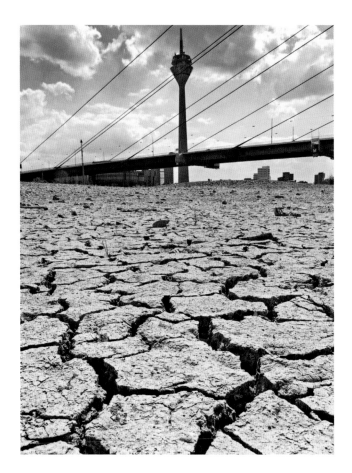

River drought The new river may be generally more
navigable, but it is also more capricious. Much of its floodplain
upstream has been drained and barricaded from the river. There are
no wetlands to act as sponges, absorbing flood flows and keeping
the river flowing during droughts. The result is more periods of high
downstream flows in cities such as Dusseldorf, where peak river flows
are today a third higher than a century ago. But there are also more
frequent episodes of low river flows, and the second picture shows
one such, during the European drought and heatwave of 2003, when
the river virtually dried up for several weeks.

ARAL SEA Kazakhstan/Uzbekistan

1973 The demise of the Aral Sea in central Asia is one of the greatest environmental disasters of the 20th century. Back in the 1960s, it was the fourth-largest inland sea in the world and had one of the Soviet Union's biggest fishing fleets. But at Muynak, the main Uzbek port, the sea disappeared over the horizon in the early 1970s. By the late 1980s the fish had disappeared and areas of salty desert were appearing where once there had been breaking waves. The great rivers that once filled the sea were no longer delivering water because Soviet engineers were abstracting most of their flow to irrigate endless fields of cotton. With no new water, the sea was slowly dying in the desert sun.

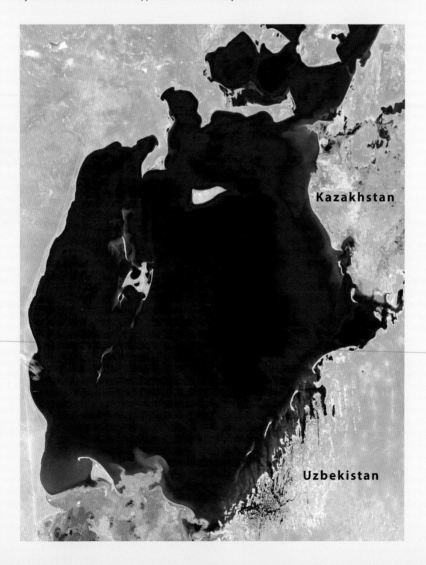

Kazakhstan

Uzbekistan

Shrinking sea Today Moscow's cotton commissars have long since departed. But the independent nations of Uzbekistan, Kazakhstan, and Turkmenistan are still emptying the rivers to grow cotton for Western markets. These three countries have the highest rates of per-capita water consumption in the world. The sea's shoreline is now 100km (60 miles) from the old seaside promenades and, for most of the time, the sea itself is split into three lakes.

Kazakhstan wants to protect one of the lakes, in the top of this picture, by damming its exit. But that will only hasten the disappearance of the other two. Meanwhile, without the moderating influence of the sea the climate of central Asia is changing. Dust storms rage, poisoning local populations with salt and chemicals picked up from the exposed sea bed.

ARAL SEA Uzbekistan

1973 Ships still trawled the waters of the Aral Sea, even as the waters began to retreat from its former shoreline. But these were the last fish harvests. With no new water being delivered from the mountains of the Hindu Kush, the sea was becoming ever more saline, until the fish died and the boats were abandoned on the dried up sea bed. A Soviet military installation is visible on an island in the distance.

A new desert The shoals of fish are today replaced by herds of cattle. The fort, long since abandoned, is now surrounded by a sea of sand. A new desert has formed across hundreds of miles of old sea bed. It is largely unexplored except by a few herdsmen who brave the new terra incognita. One day perhaps, unless the cotton kings of Uzbekistan and Kazakhstan relent and allow the water back, someone will declare the desert a protected area.

MALDIVES Indian Ocean

1995 The Maldives is a nation of 1,200 coral islands in the Indian Ocean, southwest of India. Many of the islands are unnamed and uninhabited specks of land. Nowhere on this particular island, in the centre of the archipelago near the capital Male, is more than 100m (300ft) from the shore, nor more than a few tens of centimetres above sea level. But such islands are far from worthless. Tourism is the country's biggest industry and visitors will spend large amounts of money to find their own undisturbed tropical paradise for a day, and to snorkel among the coral.

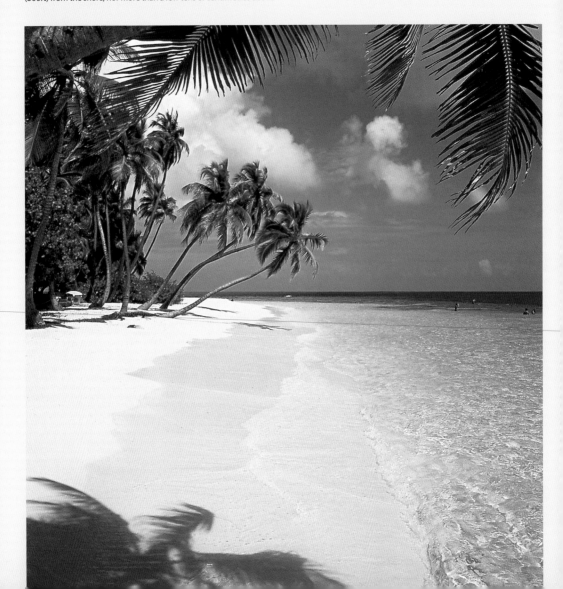

Rising tides

Sea levels are rising and the beaches on these islands are eroding fast. This is the same shore on the same island less than a decade later. The sand is largely gone, dead coral has been exposed, and the palm trees are starting to capsize into the ocean. Many of the islands of the Maldives are so low that they were briefly inundated by the Boxing Day tsunami in 2004. Few lives were lost, but the scenes provided a fearful foretaste of the islands' likely future. As sea levels rise further, most of them could be permanently underwater by the end of the century.

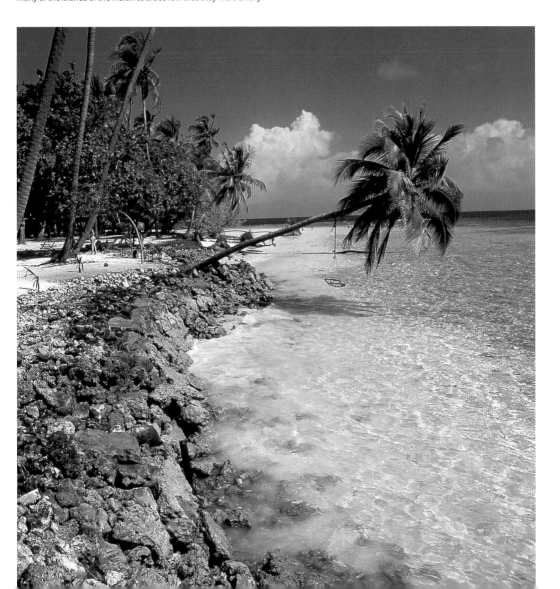

CARYSFORT REEF, FLORIDA KEYS USA

1975 Until the late 1970s Carysfort Reef on Key Largo was the largest and most luxuriant coral reef among the islands that make up Florida Keys. It was not exactly undisturbed by humans. The reef is named after a famous ship, the British ship HMS *Carysford*, which was wrecked there in 1770. No fewer than 63 ships came to grief here in the treacherous waters around the reef in the 1830s, before engineers sunk piles into the reef to install a lighthouse – the first reef lighthouse in the world. But it had large areas of pristine shallow reef containing distinctive corals such as Elkhorn and Staghorn. And divers loved its steep "cliff", which contained a huge variety of coral formations.

Dying reef Carysfort Reef is now part of the Florida Keys National Marine Sanctuary. But that hasn't saved it. The decline in the ecology of the reef since the 1970s is pictured here by the same marine biologist who took the first image. Shortly after the first picture was taken, the reef was hit by two coral diseases: White Plague and Black Band Disease. Meanwhile, sediment from pollution and other human activities fed a smothering growth of algae. And there has been increasing physical destruction of the coral by boats. Carysfort Reef is still a popular diving destination today, but it has lost 90 percent of its coral since the 1970s, and is probably in terminal decline.

TIANANMEN SQUARE, BEIJING China

1997 Tiananmen Square, in front of the ancient Forbidden City in Beijing, is one of the greatest public squares on Earth. Its 40 hectares (100 acres) are famous for many things. It was here that students called for democracy in 1919 and again, when they were brutally crushed, in 1989. The Cultural Revolution began when a million young Red Guards rallied here in 1966, and the Revolution's instigator, Chairman Mao, lies here in a marble mausoleum. On a bright spring morning the square seems to sum up the best and worst about China and its capital city.

Sandstorms China has growing ecological problems, and they, too, are visited on Tiananmen. The country's 1.3 billion people and their breakneck dash for economic growth are placing huge strains on the environment. Fertile fields and pastures thousands of kilometres from Beijing are turning to dust as the country tries to feed its people. And, when the wind is in the right direction, those crumbling fields make their presence felt in the capital as huge sandstorms rage out of the west. This is Tiananmen Square in 2003, with the Forbidden City shrouded in sand. The storm will head out across Korea and Japan, cooling the Earth's surface and dosing marine ecosystems with minerals before giving a sandy dusting to the mountains of Canada.

OZONE LAYER Antarctica

2006 The ozone layer surrounds the entire planet in the stratosphere. It protects life on Earth by blocking out harmful ultraviolet rays from the sun. Without it there would be many more cancers, cataracts, and other diseases, and ocean life could be destroyed. But since the late 1970s a "hole" has opened up in the ozone layer above Antarctica each southern spring. The hole is caused by very stable man-made chemicals that slowly rise up into the stratosphere, where they release chlorine and bromine. During the long, dark, and uniquely cold winters over Antarctica, a "soup" of ozone-destroying chemicals builds up. As soon as the sun hits the atmosphere again, it triggers a runaway chemical reaction in which small amounts of the gases can destroy large amounts of ozone. In this image, from September 2006, the purple area shows where more than 75 percent of the ozone has been destroyed – and at certain elevations the figure is more than 90 percent.

The hole can heal This is the ozone layer three months later. It has largely healed. As the spring sun gradually warms the air, the runaway destruction of ozone stops and other reactions start to re-form more ozone – until the next year. Global controls on ozone-destroying chemicals such as CFCs mean that their concentrations have now stabilized in the atmosphere. Soon they should start declining. When that happens the annual Antarctic ozone hole should gradually diminish. In perhaps 50 years, the ozone layer may resume normal service and this picture could be typical of September, too.

Mumbai is one of the world's five largest cities, with slums amid high-rise opulence. The city has the world's first $2-billion home, but also the notorious Dharavi slum, setting for the movie *Slumdog Millionaire*. Dharavi has 600,000 inhabitants in an area half the size of New York's Central Park, and still finds room for 10,000 workshops where many of its inhabitants earn a living. They even do visits for tourists.

Urbanization

For the first time in history, most humans live in cities. Urban areas still occupy only two percent of the Earth's land surface, but they house more than half of us. The unprecedented migrations of people from the countryside mean that virtually all of the world's population growth is now in cities. Around 1940 New York became the world's first megacity with a population of more than 10 million. Today there are 20 of these urban monsters; Tokyo now has 27 million people; São Paulo and Mexico City are not far behind.

Great cities and ancient trading ports such as Beijing, Dubai, and Singapore have grown, but others, such as São Paulo, have emerged from obscurity to dominate their countries. Still others, including Las Vegas in the desert of the American West, have been created in just a few decades from blank places on the map.

Cities are the engines of economic growth. They grow around activities best carried out centrally, such as government and finance, manufacturing, wholesaling, and running ports. They become centres of skilled labour and sure markets for goods, so that even when their old industries subside, new ones are able to take their place.

Cities often grow despite being noxious places, centres of disease and pollution. Tens of millions of Chinese peasants flock to cities every year, even though half a million Chinese die each year from lung disease associated with urban air pollution. See here the visual evidence of how Kathmandu, once a Himalayan idyll, is now choking on traffic fumes accumulating in the thin mountain air. And cities can be violent. The biggest cause of death among young people in São Paulo is homicide, followed by traffic accidents.

But, let's face it, people like living in cities. The streets may not be paved with gold, as Dick Whittington had hoped when he headed for medieval London, but at least they are paved. Even for the poor, cities represent the good life and the bright lights, a fact graphically illustrated by our pictures of Los Angeles. But cities also have a tremendous capacity to reinvent and improve themselves. Look at the transformation of Bilbao from an old Basque centre of heavy industry to a beacon of 21st-century culture; or the reinvention of the river through Seoul; or the contrasting regeneration of rundown Buenos Aires and the British military port of Gibraltar. See, too, how the old West India Dock in London, once the world's greatest imperial commodities port, has become Canary Wharf, one of the great global centres for trading in money and information, and how Seattle has gone from smokestacks and shanty towns to Microsoft and *Frasier*.

The products (and pollution) from cities extend far beyond their borders. But so, too, do their demands. Cities consume three-quarters of the resources we take from the Earth. Maintaining London is reckoned to require a land area 120 times that of the city itself. So look elsewhere in this book for evidence of how the demands of the world's cities impact on the rainforests of Brazil, the wheat prairies of North America, the cotton fields of central Asia, the copper mines of Chile, and, ultimately, on the ice sheets of the polar regions. The urban footprint in the 21st century is truly global.

HONG KONG China

1880 The barren Chinese island of Hong Kong was a neglected
fishing port with a population of just 7,000 until it was taken over by
the British in 1841 after the Opium Wars with China. By 1880 the
sheltered waters between the island and the mainland had become
a major port, a bustling centre for east–west trade, and a centre for
banking and insurance for the British Empire throughout Asia. Docks,
hotels, and warehouses dominated the island's narrow coastal strip,
seen in the foreground, while ferries crossed constantly to the
Kowloon peninsula on the mainland, seen here in the distance.
By 1880 Hong Kong's population was approaching 160,000.

Rapid expansion

Hong Kong reverted to Chinese control in 1997. It has long since outgrown its waterside origins, spreading upward into the mountainous centre of the island, as well as outward across Kowloon and the New Territories beyond. Its population of more than seven million now occupies one of the most densely packed conurbations in the world, with more than 6,000 people on every square kilometre (0.4sq mile). High-rise buildings dominate the island's coastal strip, the precipitous lower slopes of the central mountain, and the artificial land created by developers at the water's edge. The tallest building, pushing above the skyline at the centre-left of the photo, is the 88-floor International Finance Centre. Hong Kong and the neighbouring Chinese province Guandong now seem set to merge, forming the largest urban agglomeration on the planet.

CHONGQING China

1984 This was the heart of Chongqing, a sleepy crossing point on the River Yangtze in Sichuan province in southwest China a quarter of a century ago. Chongqing at that time had a population of around two million, spread along the banks of the river. But the narrow streets were for kids to play in, for *al fresco* cooking the tiny homes, and for hanging the washing across. The occasional bicycle, maybe. But cars? None at all.

A megacity is born

This is the same spot today, according to the photographer. Though you could not possibly tell. Urban Chinese today are among the world's most assiduous consumers, and this brightly lit superstore is just one emporium to satisfy their demands. Chongqing stands at the head of the reservoir that powers the mighty Three Gorges Dam, the world's largest single source of electricity. The city and the surrounding district are home to more than 30 million inhabitants in a sprawling megacity the size of Belgium. It is the hub of China's efforts to industrialize its vast, and still largely rural, western region. It is a major manufacturer of iron, steel, aluminium, and cars. It even has a Walmart. Attracted by the jobs and the bright lights, an extra half-a-million people move into the city each year.

DUBAI United Arab Emirates

2006 A new business district rises amid what is still
recognizably desert in the Gulf emirate of Dubai. The old port,
a centre for smuggling the world's gold and much else between
East and West, is changing. Thanks to oil money from neighbours
such as Abu Dhabi, a construction boom is under way. On the left
in the middle distance, just one of a number of half-completed
high-rise buildings is the planned jewel in the new crown, the
Burj Dubai building.

A long way down

By December 2009, the Burj Dubai is completed – and renamed the Burj Khalifa after the president of the United Arab Emirates, Khalifa bin Zayed Al Nahyan. At 828m (2,716ft) high, it is the tallest ever man-made structure. Eventually 25,000 people will live here, with a swimming pool on the 76th floor and a mosque on the 158th floor. The $1.5 billion tower is the centrepiece of one of the most dramatic real-estate booms in history, part of a $20-billion complex of buildings in downtown Dubai. But even before the tower officially opened in January 2010, Dubai was hit by massive financial difficulties in the wake of the global credit crunch, bringing a bail-out from Khalifa. Property prices crashed, sparking fears of a debt default. The profligacy of the tower was also called into question. Environmentalists pointed out that every time the toilet is flushed "they've got to pump water half a mile into the sky".

MEXICO CITY Mexico

1986 With a population of 20 million, Mexico City is one of the world's largest cities. It also used to be the smoggiest city in the world. The surrounding mountains trapped industrial pollution and traffic fumes from ill-maintained buses, taxis, and cars. This is also one of the world's highest major cities, and the thin air at 3,000m (10,000ft) slows the atmosphere's natural cleansing processes. As a result, for much of the year the mountains were barely visible. Legend had it that breathing the foul air for a day was as dangerous as smoking a packet of cigarettes. The city's reputation became so bad that international corporations were pulling out for fear of damaging the health of their executives.

Cleaner air
Since the late 1980s the city authorities have worked hard to clean up the smogs. They banned high-sulphur oil from factories and cars, imposed checks on vehicles every six months to ensure clean running, and restricted traffic during the heaviest pollution episodes. Nobody would say the smogs are banished, but some days, as here, the mountains can be seen clearly. And, in a city with ever more people and cars, at least things haven't got worse.

SAN FRANCISCO USA

1906 At the start of the 20th century San Francisco was America's smartest, newest, and most go-getting city. With some 400,000 inhabitants, it was home to a quarter of the fast-expanding population of Americans west of the Rocky Mountains. But just before dawn on 18 April 1906 the northern end of California's now-notorious San Andreas Fault shifted. In the resulting earthquake the city's buildings tumbled and fires spread through the wooden buildings. The fires did most damage, devastating most of the city – as this image looking down California Street a month later shows. The Grace Church stands out amid the ruins. The earthquake killed some 3,000 people and left most of the city's population homeless.

Rebuilding programme San Francisco, of course,
was built afresh. It was laid out in a grand manner, with wide arterial
roads through the once-ramshackle city centre and a subway system.
Reconstruction was largely completed within a decade. The buildings
in the foreground in today's photo, taken from the same spot as the
1906 image, come mostly from this phase. But in the past 50 years
a whole new central business district of high-rise blocks has risen
downtown. Today, the Ritz Carlton Hotel stands on the site of the
former Grace Church. The church itself was eventually rebuilt as
Grace Cathedral in the suburb of Nob Hill.

SEATTLE USA

1937 The 1930s was the decade when the American dream almost fell apart. Millions of people attracted to the cities by jobs found themselves out of work during the Great Depression. Unable to pay rent, many were thrown out of their homes as well. In many cities they took refuge in self-built hovels on spare land, such as here at Hooverville close to the port of Seattle. The people here had built the downtown art deco towers in the distance, and queued for jobs in the city's belching factories. But now they were unwanted and virtually under siege. City authorities several times tried to tear down Hooverville, which became a national symbol of resistance to unbridled capitalism.

Slicker city Today Hooverville is long gone, replaced by a giant container yard for the port. The descendants of its inhabitants now mostly have jobs, and homes in the suburbs. Heavy industry, too, has been banished, and the city air is much cleaner as a result. Seattle is now the city of Boeing, Microsoft, and *Frasier*, and its downtown high-rise has grown ever higher. But some 1930s buildings remain: notably the tall building from the top right of the first picture, which now appears as a midget in the shadow of taller newcomers.

SANTA CRUZ Bolivia

1975 Santa Cruz de la Sierra, in the bottom left, is Bolivia's second-largest city. It sits on the western edge of a vast area of lowland forest, bush, and wetland that once stretched almost uninterrupted to the Atlantic Ocean, 2,000km (1,200 miles) to the east. As can be seen here, as late as the mid-1970s the forest east of the River Grande was largely empty. But around then, the government decided to move tens of thousands of people from the poor highland regions of the Andes to clear the forested lowlands and plant crops in the fertile forest soils.

Urban growth
Today the forest is gone – clean-cut to make way for a huge development scheme called Tierras Baja. That scheme now covers most of the area and is connected to the outside world through a new network of roads. Migrants live in village communities, each with a church, school, bar, and soccer field. The rectangular light-coloured areas seen here are fields of soybeans, which is the most valuable and widespread crop in the region. Most soybeans are exported to Europe and North America.

TOKYO Japan

1925 In the Land of the Rising Sun, Ginza always rises again. For more than a century it has been Tokyo's premier shopping district. It was first wrecked in 1872 when a massive fire ripped through its wooden buildings, after which the government hired an English architect called Thomas Waters to rebuild it in Western style, this time with bricks. Then in 1923 came the Great Kanto Earthquake, which levelled almost every building. But this picture shows how, just two years later, Ginza was well on the road to recovery. Yet larger buildings had emerged, including the famous Waco department store at centre-left, and the street is bumper-to-bumper with trams.

Rising power Ginza lost its buzz during the austerity of World War II. And its buildings were destroyed again one night in early 1945 when Allied bombing created a firestorm through Tokyo that killed an estimated 100,000 people – more, even, than the atomic bomb at Nagasaki. But, undaunted, Ginza rose again as a brightly lit consumer paradise, the epitome of modern Japan, and home of every electronic gadget imaginable – and the $10 coffee. This time it is built of steel and glass. But will it prove more durable?

BUENOS AIRES Argentina

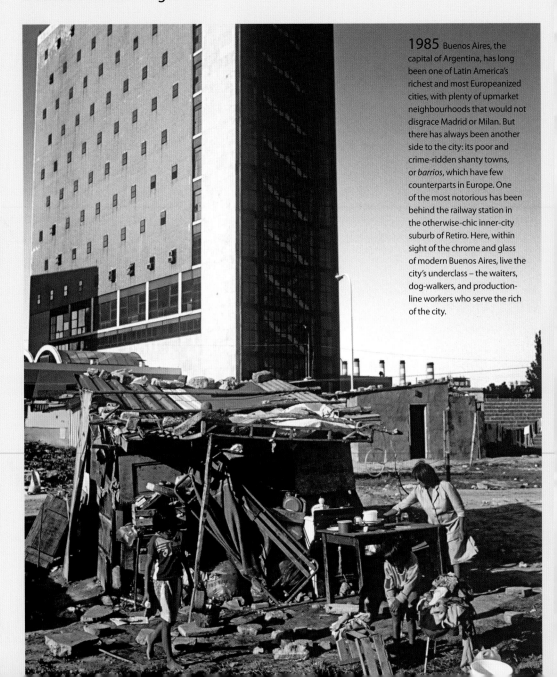

1985 Buenos Aires, the capital of Argentina, has long been one of Latin America's richest and most Europeanized cities, with plenty of upmarket neighbourhoods that would not disgrace Madrid or Milan. But there has always been another side to the city: its poor and crime-ridden shanty towns, or *barrios*, which have few counterparts in Europe. One of the most notorious has been behind the railway station in the otherwise-chic inner-city suburb of Retiro. Here, within sight of the chrome and glass of modern Buenos Aires, live the city's underclass – the waiters, dog-walkers, and production-line workers who serve the rich of the city.

Self-help scheme

It was only in the 1990s that paved streets and basic services such as regular electricity and clean water arrived behind the railway station. But with the arrival of those services the *barrios'* residents have begun to construct a better community. However, self-help is the order of the day. Twenty years after the previous photo was taken, photographer Mark Edwards returned to the same spot and found that the new residents were replacing the thin plywood and corrugated iron shack with a new home of bricks and concrete. Retiro, at long last, is on the up.

MILLAU VIADUCT France

2001 The Tarn Gorge had long been a traffic bottleneck for drivers on the *Route Nationale* N9 near Millau in southwest France. It is one of the main routes from Paris to Languedoc and Spain. Every summer, traffic built up as the road zigzagged down into the 200m- (650ft-) deep gorge, across a small bridge over the river, and back up again. The gorge was such an environmental blackspot that even green activists were generally in favour when the government announced plans for a bridge to span the valley.

Seven-pier wonder After four years' construction the Millau Viaduct opened in December 2004. It stretches for 2.5km (1.5 miles) across the top of the valley and is the tallest road bridge in the world. One of its seven piers soars higher above the valley bottom than the Eiffel Tower does above Paris. Designed by British architect Sir Norman Foster, the bridge has been hailed as a design marvel, and a platform for viewing the bridge has become a major tourist attraction in its own right.

SPROGO ISLAND Denmark

1988 Sprogo is a small island in the middle of the Storebaelt, or Great Belt – one of two major natural waterways linking the Baltic Sea to the North Sea. It has a lighthouse, a cottage for the lighthouse keeper, a jetty, some rudimentary sea walls, a few trees, some intriguing Neolithic remains, and that's about it. On the map it looks like a stepping stone between the larger Danish islands of Funen and Zealand, the latter of which houses the Danish capital, Copenhagen. And that's how things have turned out.

Extended span In the 1990s engineers constructed the 17km (9-mile) Storebaelt link between Zealand and Funen, via Sprogo. The western bridge in the distance, between Funen and Sprogo, has two decks carrying a motorway and a railway. Stretching east toward the camera on the left is the start of a 7km (4.5-mile) suspension road bridge from Sprogo to Zealand. And on the right is the start of a railway tunnel to Zealand. In the process of creating this mid-ocean transport hub, land reclamation has quadrupled Sprogo's size. The project forms part of a grander scheme for a series of giant transport links to span several entrances to the Baltic Sea, providing a direct link to northwest Europe for the first time.

VALE OF NEATH Wales

1994 The valley of the River Neath was one of south Wales' most peaceful, with woodlands, fields, a handful of small villages, and the Glyncorrwg hills behind. But transport engineers had had their eyes on the valley for a long time. It was earmarked to form a vital link in the planned A465 "heads of valleys" route through the former mining regions, from Abergavenny to Swansea, and become a new part of Europe's trunk-road network. And, in 1994, the planners got their way.

Major road link This is the same scene three years later. The £80-million Aberdulais to Glynneath link road is complete. Fields have been turned to concrete and the rural tranquillity is gone for good. The valley, once most famous for its walking trails to the Aberdulais Falls, is now zoned for economic development. Some want to see industry return to a valley that used to be home to miners; others foresee a new "industrial heritage industry" based on the Neath Canal.

BEIJING China

1978 In the last days of China's revolutionary communist leader Mao Ze Dong, Beijing was a relatively austere and tightly packed capital. It forms the small grey area in this satellite image. The hills to the west are still green with forests, and farms show as red, orange, or yellow, depending on the crop being grown. Chairman Mao's era had lionized the rural peasant and there were strict controls on people moving to the cities, where, in any case, all jobs were under state control. Economically, the capital of the world's largest country, already with around a billion inhabitants, slumbered.

City migrants Economic reforms begun in 1979, three years after Mao's death, unleashed private capitalism on China, albeit under state control. The country began more than two decades of rapid economic growth, with GDP increasing by about 10 percent a year. China's population growth was capped by a rigorous birth control policy, but as the economy boomed, millions of migrants flooded from the countryside into Beijing and other cities. This second image shows how the city's suburbs have grown to house the migrants, taking over huge swathes of former farmland and engulfing market towns. The city's population has almost doubled in size to 15 million.

AVENUE PAULISTA, SÃO PAULO Brazil

1902 In the late 19th century São Paulo was a Brazilian backwater, smaller and less famous than Rio de Janeiro down the coast. Avenue Paulista was one of its major thoroughfares. By the turn of the century, when this picture of the avenue was taken, the state's coffee industry was booming, bringing in millions of migrants from Europe to work the plantations. Many of the estate owners lived on the avenue, which remained a relaxed, spacious, and tree-lined retreat from the bustle of global trade.

Megacity

Avenue Paulista today is no retreat – it is the hub of one of the world's great megacities. With a population of some 19 million, São Paulo's metropolitan area ranks only behind Tokyo and Mexico City. The city's coffee business, though still vibrant, has been overtaken by many others, from steel to car manufacturing and textiles to electronics. And many of the finance houses and corporate headquarters that are making Brazil's economic growth second only to that of China are on Avenue Paulista. Not surprisingly, it has some of the most expensive real estate in Latin America. But down there, somewhere, a few of the original mansion houses remain as reminders of a more leisurely past.

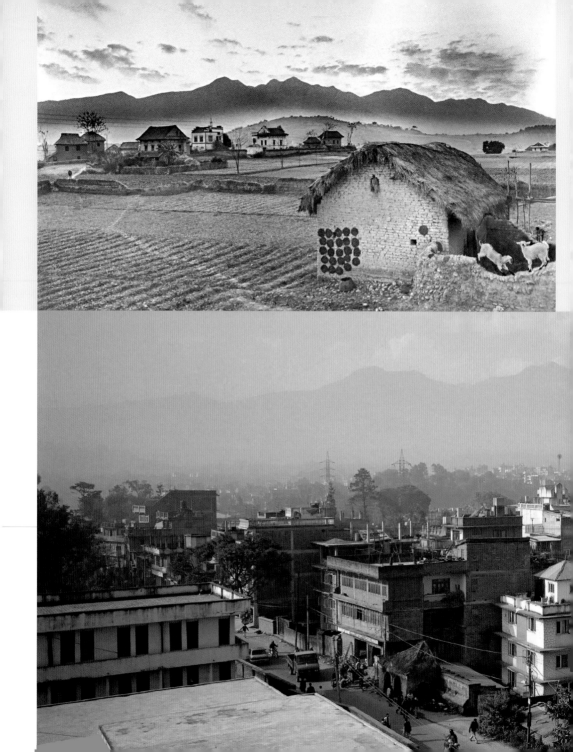

KATHMANDU Nepal

1970 High in the Himalayas, Nepal is a land of goats and small fields, of clear air and rural communities connected only by long footpaths through the mountains. In 1970 virtually the entire population of Nepal lived in the countryside, much of it extremely remote. This is a scene 1,300m (4,300ft) up in the Kathmandu valley, several kilometres from the nation's small administrative capital of the same name.

Urban sprawl The lower shot shows the same spot today, and was taken by the same photographer. The houses, the village, and the fields are all gone – the sprawling suburbs of Kathmandu have consumed them all. A city of only 200,000 people in 1970, having engulfed its neighbours Patan and Bhaktapur and most of the land in between, it now has a population of more than 1.5 million. Unlike in many other Asian countries, urbanization in Nepal has not been accompanied by rising living standards. The country remains one of the poorest, least industrialized in Asia. People have come to Kathmandu to escape grinding poverty and in the hope of finding basic services such as education and health care. But at this spot, villagers have become city-dwellers without moving. Of the former landscape only the mountains remain – though they are so shrouded in the smog accumulating in the rarefied air that they are barely visible.

GIBRALTAR Mediterranean Sea

1954 The Rock of Gibraltar occupies the southernmost tip of the Iberian peninsula. Though claimed by Spain, its big neighbour, the 6.5sq km (2.5sq miles) of Gibraltar have been a British dependency since the Treaty of Utrecht in 1713. The harbour, which commands the entrance from the Atlantic Ocean to the Mediterranean Sea and the Suez Canal, has been occupied continuously by the Royal Navy. Though lacking the grandeur of Asian city-states such as Singapore and Hong Kong, Gibraltar has tried to lead an independent existence – catching most of its water from tanks set into the Rock, for instance. Half a century ago, however, it remained little more than a naval outpost, with meagre housing creeping up the slopes of the Rock.

Steady growth Spain continues to block access to the airport for flights other than to Britain, but otherwise relations have improved now that both Spain and Britain are in the European Union. The dockyard accounts for only seven percent of the dependency's economic activity, compared with 60 percent as recently as 1984. And modest prosperity has come to the Rock, albeit sometimes due to its convenience as a tax haven, a centre for online gambling, and as a staging post for drug-running from North Africa into Europe. Certainly the marina is flourishing, and high-rise buildings have grown along the shore.

CANARY WHARF, LONDON England

1976 London's West India Docks were among the world's largest when they opened in 1802 – unloading produce such as rum, sugar, and molasses shipped from British colonies in the West Indies. The docks occupied most of the Isle of Dogs and were connected to the River Thames by giant locks. They could berth 600 ships at a time, decanting their contents into more than 1 km (0.6 mile) of brick warehouses. But German bombing during World War II destroyed most of the warehouses, and by the time this picture was taken most of the business had departed for bigger ports, such as Felixstowe, that could handle containers. The dock finally closed in 1980.

Urban regeneration

After being derelict for less than a decade, the Isle of Dogs was reborn. The old hub of Victorian imperial trade has been replaced by an axis of 21st-century global finance. Canary Wharf, named after the part of the West India Docks that handled trade to the Canary Islands, is the heart of the regeneration of London's Docklands. It contains Britain's three tallest buildings, employs more than 80,000 people, and houses major banks, law firms, and two national newspapers. Already a rival to London's traditional financial centre, "the City", 6km (3.7 miles) upstream, there are plans to double its size. Behind it, on the opposite bank of the Thames, is the distinctive white shape of the Millennium Dome, renamed the O2 Arena – a symbol of modern London.

SEOUL South Korea

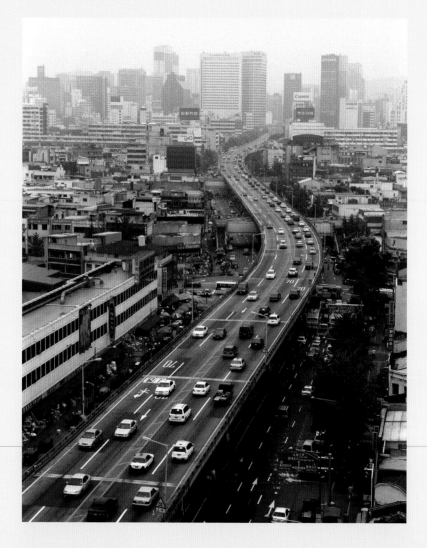

2003 Through the 1960s and 1970s Seoul, the South Korean capital, was one of the fastest-growing cities on Earth. Effluent discharges reduced its main river, the Cheonggyecheon, to little more than a sewer. So the government completed the job by encasing the river in concrete and building a six-lane highway above it. From London to Tokyo, many urban rivers have been killed off in this way – a final act to banish nature from the urban environment.

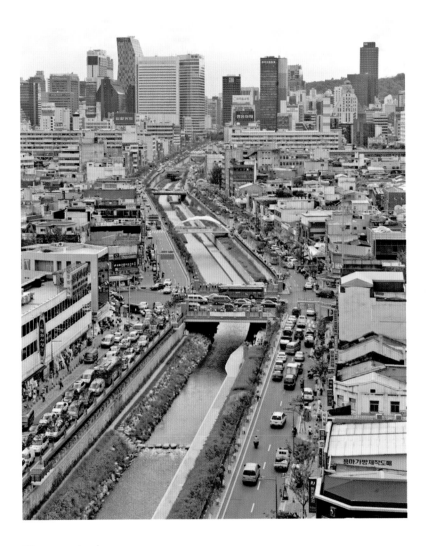

River revival But in Seoul the mayor wanted to turn the tide. In 2002 he decided to get rid of the road, restore the river, and create an 8km (5-mile) park along its banks. Despite widespread opposition, the restoration has been achieved. Drivers were offered an improved bus service. Besides improving the city's landscape and providing a centre for recreation and wildlife, the revived river also acts as an air conditioner for the city. On its banks summer temperatures are more than 3˚C (5.4˚F) lower than when it was concrete. Other cities, including Shanghai and Los Angeles, are now following Seoul's example.

DOWNTOWN Singapore

1962 Singapore has for more than a century been one of the great trading ports of the world. In 1962, in the midst of a messy divorce from Malaysia, the 50km- (30-mile-) long island state housed some 1.6 million people. Three stages in its development are shown here: the rowing boats and sampans of its Malay Chinese nautical origins; the old colonials' shorefront buildings from the height of the British Empire; and the 1950s building to the left, the Asian Insurance Building, which was at the time the city's tallest.

Upward mobility
Since breaking with Malaysia, independent Singapore has grown rich as a port and financial centre to rival London, New York, and Tokyo. It became the smallest but most dynamic of the "tiger economies" of Asia. Since the 1960s it has grown vertically to accommodate a population that now numbers 4.5 million. Most remarkably, notice the Asian Insurance Building in this modern picture. Nestled close to the far left of the picture, the building that was once the city's largest is now the smallest thing in the central business district.

BILBAO Spain

1948 Bilbao has long been one of the "smokestack" cities of Spain. Feeding on the iron reserves in nearby hills, it became a centre of the steel and shipbuilding industries in the heart of the Basque country, and for a long time the richest city in Spain. But wealth came at a cost. Fire and smoke often filled the air, as in this dramatic post-war industrial landscape in the dock area of Euskalduna and close to the Basconia steelyard.

Cultural shift Almost half a century on, Euskalduna has been transformed into a centre for services and culture. Like many centres of heavy industry in Europe and North America, it shed its former economy activities almost entirely in the 1980s. With the shipyards and steel mills gone, Euskalduna is now a waterfront development of parks, apartments, shops, and offices. And this totemic piece of architecture – the Guggenheim Museum, which opened in 1997 and is dedicated to 20th-century art – symbolizes the change. Only the ship shape seems vaguely familiar.

LAS VEGAS USA

1973 "Vegas" was founded in 1905 as a desert railway town used by the silver and gold miners of Nevada. For much of the time since, it has been the fastest-growing urban area in the USA. First, it was a base for construction on the nearby River Colorado of the Hoover Dam, the world's first superdam. And since the 1940s it has developed a new identity as a centre for hotel-based casinos, entertainment, and conventions, fuelled partly by the nationwide move from the north and east to the "sunbelt" states of the south and west. At the time of this image the permanent population was around a quarter-million.

Watering the strip
Today, driven by the development of resorts such as the Mirage, which opened in 1989, the city has spread across most of the available desert floor. It is approaching the Spring Mountains to the west. The population of the urban area exceeds a million, and the growth of the paved area across the desert floor has accentuated the danger from flash floods coming out of the mountains during the occasional heavy rains. And it suffers, too, from water shortages as local aquifers are pumped out to maintain the city's fountains and water its parks and palm trees.

LOS ANGELES USA

1908 Los Angeles began as a farming village on the banks of the Los Angeles River in the late 18th century. By 1908, when this picture was taken from the summit of Kitt Peak 30km (19 miles) away, it was growing fast. It had doubled its population to more than 250,000 since the start of the decade. Its street lights were still just a pool of light in one corner of the Los Angeles Basin, but the city had big plans. Film studios had just opened for business in Hollywood, and the city fathers were building a reservoir and aqueduct to tap water from 300km (186 miles) away in Owens Valley, which would underpin the future expansion of the desert city.

Light pollution Now a megacity of some 13 million people, Los Angeles and its suburbs burn bright right across the valley to the base of Kitt Peak. The city is visible for hundreds of kilometres. The searchlight on the top of one hotel can be seen some nights in Death Valley 150km (93 miles) away. But all that light consumes a lot of energy in climate change-conscious California. And the glare means that stars are no longer visible in the night sky. LA is today trying to cut its light pollution, but clearly there is a long way to go.

LONDON England

1946 Just a year after the end of the World War II, the bustling
River Thames in central London was as busy as it had been for
a century. Looking upstream from Tower Bridge and the "pool"
of London, where ships and barges congregate, to the dome of
St Paul's Cathedral, its banks are crowded with warehouses. Bar some
bomb damage during the war, the skyline is unchanged and London
remains the heart of the largest empire the world has ever seen.

A new century Here is the same view at the start of the
21st century, with one speculative addition. New landmarks are
evident, including the bulbous shape of London City Hall in the
foreground on the left. St Paul's is much less prominent in the
distance. But by far the tallest building, perched on top of London
Bridge station, is the proposed Shard, a 306m (1,004ft) glass spire
that may, or may not, one day dominate the London skyline.

Wind pioneer Altamont Pass in California is one of the world's pioneers of wind power. Its first turbines began beating in the late 1970s. They were installed after the energy crisis triggered by soaring oil prices in 1974. Wind farms in Europe now have much larger turbines and generate more power, but Altamont remains the global symbol of the world's growing love affair with renewable energy.

Land Transformation

Mankind has been exploiting the land ever since we left our first footprints on the Earth. For thousands of years we have ploughed pastures and chopped down forests, gouged minerals from the Earth and irrigated deserts. But it is only in the past two centuries or so, since the beginning of the Industrial Revolution, that our imprint on the planet has become truly indelible. And the great majority of our impacts have been within that relatively minuscule time frame that is the past half-century.

Take mining. Winning metals, minerals, and fuel from the Earth is the world's largest industry: uranium for bombs and phosphates for fertilizer; aggregate for roads and copper for cable; oil to fuel your car and coal to generate your electricity; polyester for your shirt and coltan for the capacitor in your mobile phone; bricks for your house and stones for your patio; talc for your talcum powder and diamonds for your jewellery; aluminium for aircraft and steel for almost everything. All come from the Earth.

Today the easy supplies of most of these vital resources have run out. Ever-larger amounts of rock ore have to be extracted to keep industry supplied. Production of copper has increased 22-fold in the 20th century. But the amount of ore ripped from the Earth to produce that copper has increased more than a hundredfold. Hence the huge holes in the ground where copper is mined at Bingham Canyon and Escondida.

Mining has laid waste huge areas of forest and farmland around the world, and forced the expulsion of an estimated 100 million people from their homes. On Nauru, miners have turned an entire Pacific island into a moonscape to provide nitrate fertilizer for Australian farmers. Some of the world's former mines, as we show, have been partially rehabilitated. That is good news. But bringing nature back is often hard, and has not even been attempted in most places. The scars will often still be there in millions of years.

The products of mining create pollution that fills the air in cities and eats away at great monuments. Meanwhile, forestry and farming, too, now take place on an industrial scale. This chapter shows the evidence of how some of the largest and most remote forests have been logged and their land turned into plantations or raising grounds for prawns.

In addition to these direct impacts on the terrain, huge dams now plug great rivers such as the Euphrates in Turkey, the Colorado in the USA, and the Yangtze in China. They do so in order for us to generate electricity and provide water for irrigation projects that grow cotton, wheat, and even rice in the world's dry lands. The flow of these and many other rivers is now under total human control. And we do not always use that control wisely. The Aral Sea in central Asia, once the world's fourth-largest inland sea, has been reduced to three saline sumps surrounded by a vast toxic desert.

GLEN CANYON USA

1958 The Glen Canyon, an extension of the Grand Canyon on the Colorado River, was one of America's finest natural features. But for decades US engineers eyed the Colorado as a source of both hydroelectricity and water for irrigating the arid American West. They built the Hoover Dam downstream in Boulder Canyon in the 1930s, and by the 1950s they were keen to build another.

Damming evidence

And they did. This is the Glen Canyon today. Plugged by a dam in 1964, the canyon was renamed Lake Powell – in memory of John Wesley Powell, a one-armed Civil War hero who, in 1869, made the first boat journey by a white man through the canyon. The reservoir took 17 years to fill. The Colorado River is, as a result of the Hoover and Glen Canyon dams, one of the most regulated rivers in the world. The two reservoirs can hold several years' flow, and virtually no water ever escapes from the river into the ocean. In recent years drought has lowered reservoir levels, revealing tantalizing glimpses of the old canyon – such as the white "bathtub ring" in this 2003 image. And, more than 40 years after the canyon was flooded, there is a growing campaign to break the dam, empty the reservoir, and bring back the canyon – permanently.

THREE GORGES China

1997 Since the 1930s, successive Chinese leaders have dreamed of damming the Yangtze, the world's third-largest river, where it exits a scenic mountain region known as the Three Gorges. They wanted to tame the great floods that sometimes bring havoc downriver, to make the Yangtze navigable all the way to Chongqing 1,300km (800 miles) inland, and to generate hydroelectricity that could industrialize the country's interior. Finally, in 1994, and despite growing opposition from environmentalists and people whose land would be flooded, the project got under way. Here, three years into construction, an endless line of trucks dumps earth and rock into the ever-narrowing gap across the river near the village of Sandouping.

Controversial barrier The dam is now complete.
And the 600km- (400 mile-) long reservoir behind is filling. Three Gorges is the world's most powerful hydroelectric dam. Its 26 turbines have a capacity of 18,000 megawatts. Its huge sea locks will let ships upstream to Chongqing, which is now one of the world's fastest-growing cities. But the costs have spiralled to an estimated $100 billion. Numerous archaeological sites have been flooded. And two million people have been removed from their fertile valley homes and relocated, often on poor hillsides.

ATATURK DAM Turkey

1982 The River Euphrates, which flows from top right to bottom left through this satellite image, has watered some of the world's earliest and most powerful civilizations on its journey from Turkey through Syria and Iraq to the Persian Gulf. Many of those civilizations had manipulated its flows, but nobody had attempted to halt the river's flow altogether. Not until the 1980s, when Turkey began to assemble a cascade of dams in the Anatolian highlands where the river rises and gains most of its water.

Water hoarding

Turkey's Southeast Anatolia Project is intended to develop the eastern, Kurdish half of the country by using the river to generate hydroelectricity and irrigate crops. The centrepiece of the project is the Ataturk Dam, a 184m (600ft) wall of rock and earth across the river near Urfa. It was completed in 1990 at a cost of $3 billion. The huge reservoir, seen here, now covers more than 800sq km (300sq miles) and has displaced some 40,000 people. At times during the early 1990s, while it was being filled, the dam stopped all flows downstream into Syria and Iraq. Today, the dam's operation continues to upset the river's flood regime, helping dry out the Mesopotamian marshes in southern Iraq.

MARJAH Afghanistan

1963 Most of Afghanistan is arid and barren. Farming is only possible in a sixth of the country, mostly scattered valleys where rivers from distant regions have delivered fertile silt. But even here, irrigation from diverting the rivers is usually essential before much will grow. And irrigation water also brings salt, which accumulates in the soil. Many early civilizations in this part of the world, including the city-states of Mesopotamia in modern-day Iraq, foundered when salt made their soils toxic and killed their crops. In the Marjah district of Helmand province in southwest Afghanistan, much of the irrigation and drainage infrastructure was installed by American technicians in the 1950s and 1960s to grow wheat. The area was known locally as "little America". Here an Afghan farmer from that time is digging a ditch to drain salty water from his irrigated fields.

A new crop Much has changed in Marjah since the 1960s, including the crops that farmers grow. Wheat has largely been replaced by a new and lucrative cash crop – poppies, to provide opium for the world's heroin trade. Marjah is close to towns that have made headlines during the war with the Taliban insurgency, such as Lashkar Gah. Where the government's writ does not run, poppies are the crop of choice. In 2009 an estimated 10 percent of all the world's illicit opium came from Marjah – much of it irrigated thanks to American know-how. But, whatever the crops, irrigation and drainage are essential.

PANAMA CANAL Central America

1910 The Panama Canal is one of the great engineering schemes of the 20th, or any other, century. This shipping route through Central America, from the Atlantic to the Pacific, cut 13,000km (8,000 miles) from a typical journey between the two oceans. But it was a huge endeavour. A French project to cut through the rainforest-covered mountains of Panama foundered in the 1880s with 22,000 workers dead, many from malaria and yellow fever. But US engineers took over and completed the 77km (48-mile) canal in 1914 – albeit at the cost of another 5,000 workers' lives. The most difficult task was creating the Culebra Cut – also known as the Gaillard Cut after its chief engineer – through the mountainous central spine of the isthmus. This shot was taken at Empire, an engineering base on the Cut.

Troubled thoroughfare Here is the Culebra Cut today, after recent widening to allow two large ships to pass each other. Each year more than 14,000 ships carry approaching 300 million tonnes of cargo through here. The journey takes about nine hours. But some modern ships are too large for the canal. And the canal is also beset by droughts caused, it is believed, by deforestation of the isthmus. As its reservoirs empty, there is sometimes insufficient water in the dry season to keep the Cut full.

WADI AS SIRHAN Saudi Arabia

1986 Wadi as Sirhan is a shallow depression in the vast Saudi Arabian desert, close to the country's northern border with Jordan. The depression contains fertile alluvium, but the area is so arid that it has always been a barren wasteland occupied only by passing bedouin and their camels. But hidden beneath the desert lies one of the largest underground reservoirs of water on the planet. The water mostly fell there in wetter times, and much of it is more than 20,000 years old.

Greening the desert In the past two decades, the Saudi government began to bring this underground water to the surface to irrigate crops. Each green disc in this image is a field irrigated from a rotating sprinkler arm up to 300m (1,000ft) long. Most of the fields grow wheat. But growing wheat in the Saudi desert is hopelessly uneconomic, sustained only by heavy state subsidies and the free supply of water. Worse, it is very inefficient. Much of the water evaporates before reaching crop roots. Today, thanks to such projects, more than half the country's ancient and irreplaceable underground water reserves are gone. This is much to the anger of Saudi Arabia's Jordanian neighbours, because the rocks containing the water extend either side of the border, and Jordanians believe these Saudi fields are being irrigated with water sucked from beneath their soil.

ATHABASCA VALLEY Canada

2009 The forests of northern Alberta are part of the great band of boreal forests that gird the globe south of the Arctic tundra. These forests cover even more land than tropical rainforests, and are among the largest wilderness areas remaining on the planet. But the threats to them are increasing, not just from loggers and miners, but now from oil companies intent on finding oil in the tar sands beneath the ground, particularly in this region of Canada. Here, some of the forest is still intact. But it has been staked out by prospectors for future exploitation of one of the world's last and largest remaining reserves of oil.

Tar sands
This is what happens next. This is the Suncor Millennium Mine in the once-forested Athabasca valley. The trees are gone and the soil and peat are stripped bare to reveal the oil-bearing sands beneath. Extracting the oil from the sands requires large amounts of energy, and also water pumped from local reserves and then discharged as contaminated waste into the local landscape. The oil may be valuable to Canada, but the damage to the landscape and one of the world's last great wildernesses is enormous.

ALMERIA Spain

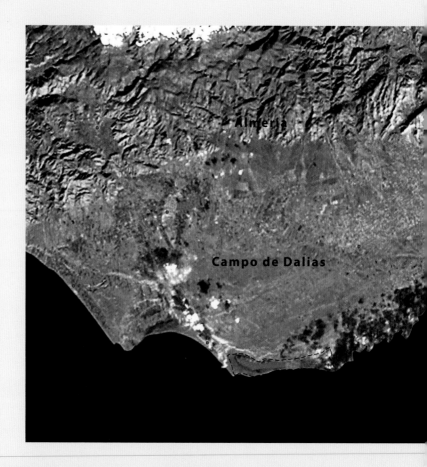

Almeria

Campo de Dalias

1974 Almeria in southern Spain is arid. In the 1970s, spaghetti westerns were filmed here. Back then, farming was patchy and small-scale, largely for want of water. This image of the Campo de Dalias, south of the mountains of the Sierra Nevada, was taken in January and shows the land fairly green, but in summer it was yellow. Soon afterward, more intense irrigated farming began to develop, as farmers pumped water from underground and state authorities tapped rivers flowing through the mountains. And in 1986 Spain joined the European Union, opening up huge markets for agricultural produce that grew best in hot sun, such as tomatoes, oranges, and citrus fruit.

Plastic plague Today intensive farming has transformed the landscape of Campo de Dalias. The yellow and green has been replaced by the white of greenhouses glinting in the July sun. The plastic maximizes temperatures and boosts output through the year. Almost every patch of agricultural land is now irrigated and under plastic. The only areas spared are the mountains and the coastal strip, which is dominated by tourist developments. But the water is running out. Local underground reserves are drying up and plans to pipe water from Spain's wetter northern areas have foundered. Time for a revival of the spaghetti western?

Almería

Campo de Dalías

PARA Brazil

1990s Mangrove forests are among the most distinctive and ecologically important types of tropical rainforest. Made up of many species of salt-water tolerant trees and shrubs, they grow in the brackish inter-tidal waters of coastlines from southeast Asia, India, and West Africa to here in Para state in eastern Brazil, near the mouth of the great Amazon River. The trees sink their roots into the silt but draw much of their sustenance from the water itself. Mangroves are vital nurseries for fish stocks, nurturing fry among their roots. They also provide physical protection from coastal erosion and hurricanes by dissipating the immense energy from winds and waves. During the 2004 Indian Ocean tsunami, areas with mangroves suffered much less damage than those where they had been removed.

Mangrove clearing

Mangroves around the world have come under assault in recent years. Over half have been lost. The land and waters they occupy is often taken for commercial farming of fish and prawns. And their timber has many uses, from fishing traps and boats to making toilet paper for sale around the world. Here, mangroves on the banks of the Amazon in Para have been cut down to make fishing traps, leaving the land behind exposed to the full fury of nature.

RONDONIA Brazil

1975 Rondonia is one of the most remote regions of the Brazilian Amazon. High in the headwaters of the great rainforest river, it is 2,000km (1,200 miles) from the Atlantic Ocean. In 1975, as this satellite image shows, the forest canopy was virtually unbroken. The few small towns in the state, such as Rio Branco, were the homes of tribal communities and small bands of Brazil nut harvesters, many of them descendants of the region's rubber boom almost a century before. Crucially, there were no all-weather roads into the region from the great population centres of Brazil to the east.

Deforestation

Deforestation A quarter-century later, the pristine canopy of the forest has been lacerated by loggers. The paving of the BU-364 highway through the province brought loggers and thousands of migrant farmers from the east. Their activity is concentrated along the dense network of sideroads created by logging companies after the highway came through. The result, seen from the air, is a distinctive herringbone pattern of deforestation. The only large surviving areas of intact forest are those protected as national parks or for indigenous tribes, including the areas in the bottom of the picture.

SUBTROPICAL FOREST South America

1973 Three decades ago this was a remote land of rainforests and rivers in the heart of Latin America, where Brazil, Argentina, and Paraguay meet. Ecologically, the most treasured area was the Iguazu National Park, a protected rainforest in the Argentinian sector in the bottom right of the picture. Three small towns – one in each country – hugged the river banks where the south-flowing River Parana met its west-flowing tributary, the Iguazu. As an ominous sign of things to come, an extensive herringbone pattern of forest clearance for farms is evident on the new roads west of Ciudad del Este.

Three of a kind

Today much of the area has been transformed. There are three giant hydroelectric dams in this picture. By far the biggest reservoir is the one behind the Itaipu Dam on the Parana, where it forms the border between Brazil and Paraguay. For 25 years, until the completion of China's Three Gorges Dam, Itaipu was the largest hydroelectric plant in the world, supplying power to Brazil's great industrial centres of Rio de Janeiro and São Paulo, as well as to Paraguay. At the top left is Paraguay's Acaray reservoir and, at the bottom right, Argentina's Urugua-i Dam. Meanwhile, the three towns have grown hugely. Ciudad del Este is now Paraguay's second-largest city with a population of a quarter of a million, mainly Asian migrants. And, while much of the Argentinian park survives, most of the Paraguayan forest has been converted to farming.

BINGHAM CANYON USA

1964 The Bingham Canyon copper mine near Salt Lake City in Utah was opened in the early 1960s. Miners extracted the copper in layers, or benches, that they blasted in turn. The copper-containing rocks, known as ore, were loaded by giant shovels into electric train cars that ran on rails beside each bench. The ore was taken away for refining and smelting to create pure copper, one of the world's most widely used metals.

Great excavations Bingham Canyon is today owned by the mining giant Rio Tinto. It is the world's largest man-made hole. With dozens of benches blasted into the Earth, it is now 3km (2 miles) across and more than 1km (0.6 mile) deep. Today's electric shovels can lift 100 tonnes of ore in a single scoop, and the dump trucks that have replaced the rail cars can carry 250 tonnes. Half a million tonnes of rock is cut daily from the mine, which has so far produced 17 million tonnes of copper, as well as by-products of gold and silver. The mine is likely to stay open until at least 2013.

ATACAMA DESERT Chile

1989 The Escondida Mine in Chile's Atacama desert is the world's most productive open-pit copper mine. It produces nearly 10 percent of the entire world's supply of the metal, as well as smaller amounts of gold and silver, and is owned by two of the world's largest mining companies, BHP Billiton and Rio Tinto. This aerial image was taken shortly before the first output from the new mine. It shows the mine itself, roads to the site (which is 3,000m/10,000ft above sea level), and, in the bottom left corner, the site of an intended reservoir for liquid waste from the metal-refining processes.

Digging deep The mine has been growing continuously. More than 120,000 tonnes of ore are removed from it every day. The ore is milled onsite to produce a copper concentrate that is mixed with water to make a slurry that can flow down a 170km (106-mile) underground pipe to the port of Coloso, where it undergoes further processing. The tailings reservoir now contains some three billion tonnes of waste sludge. It has recently been redeveloped to recycle water, the most precious resource at the desert mine. The mine's output is currently approaching a million tonnes of pure copper a year, which is used for everything from copper cables to plumbing to mobile phones.

BOSTON USA

1987 Like most US cities, Boston was dominated by the automobile. Having been laid out long before the car was invented, its tangled streets were saved from gridlock only by the creation of an elevated highway through the heart of the central district. The artery was widely disliked, however, being dangerous, unwieldy, and ugly, and because it cut off downtown from the harbour area.

The Big Dig The city fathers acted. The Central Artery/Tunnel Project, known to Bostonians as the Big Dig, put the central artery into a 5.5km (3.5-mile) tunnel dug beneath the city. Here is the new-look Atlantic Avenue. Its skyline is dominated as before by the old Customs House, but in place of the elevated highway is the Rose Kennedy Greenway. The Big Dig, completed in 2007, didn't come cheap. It is estimated to have cost the city $22 billion, making it the most expensive highway project ever in the US. The bills won't be paid off until 2038.

NAURU Pacific Ocean

1968 The tiny Pacific island of Nauru was covered in a thick layer of phosphate rock, the dried remains of thousands of years of bird droppings. Its only industrial activity for the past century was mining that phosphate to provide fertilizer for farms in Australia and elsewhere. So single-minded was the mining that by 1968, the year the island won independence, 17 of its 21sq km (8sq miles) of land looked like this: a moonscape of limestone pinnacles left behind after the removal of the island's only natural resource. The cranes sit on a small area not yet excavated.

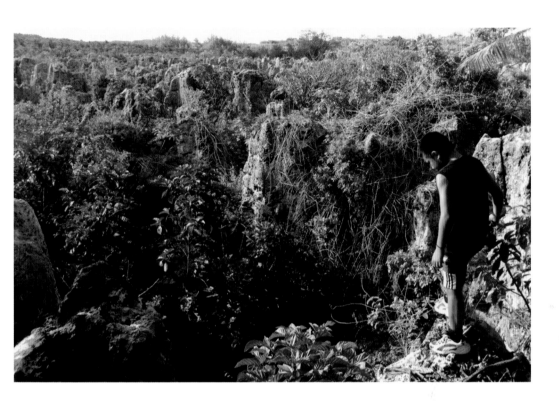

Mined out
The mining has all but ceased today. The
phosphate is largely gone and vegetation is slowly returning to the
moonscape. But after decades of providing fertilizer for the world's
farms, the 13,000 inhabitants of Nauru have no soil themselves
on which to farm. A £50-million payment from Australia in the 1990s
to compensate for the destruction of their island made Nauruans
among the richest in the world, per head of population. But the
money has been frittered away in bad investments; Air Nauru's
last Boeing was repossessed in 2005. Since 2001 Nauru has accepted
Australian aid in return for being used as an offshore detention centre
for asylum seekers. The phosphate island is now little more than an
offshore prison hulk.

VANCOUVER Canada

1907 Early settlers to the Canadian Pacific coast called it the little mountain, a tree-covered granite outcrop in the midst of the flat plains that became Vancouver. But as the city grew, the 170m (560ft) forested hill became a vital resource. First it was logged for timber, and then at the start of the 20th century miners opened three quarries at its summit to provide crushed granite for construction of the first paved roads in the booming sawmill town. This picture was taken at the height of the quarrying era. But three years later the granite was exhausted and the quarries were closed.

The return of nature
As Vancouver continued to grow, it encircled the pockmarked and deforested mountain. So the city fathers decided to create a park on its slopes, which they planted with all the native trees of Canada. Queen Elizabeth Park was opened in 1939 by King George VI and his consort Queen Elizabeth (the mother of the current Queen of England). At its centre they turned one of the pits into a sunken garden, seen here. The quarry gardens are now one of the top tourist attractions in Canada's third-largest city, drawing six million visitors a year.

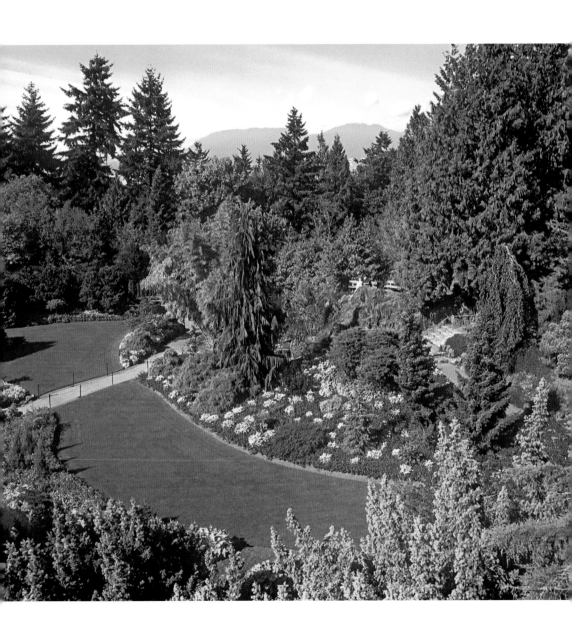

ISAHAYA BAY Japan

1993 Japan's civil engineers like land reclamation projects almost as much as the Dutch. They say that their mountainous but densely populated country needs as much extra land as it can get. Isahaya Bay on the island of Kyushu was a prime candidate. The bay is full of tidal flats that are barely flooded. In fact, it contains Japan's largest area of tidal flats. As the 1993 image shows, farmland encroached right up to the shoreline, constrained by sea walls but seemingly eager to advance farther. Small-scale land reclamation, launched in 1989, was already under way at this point.

Land reclamation

Major reclamation began in Isahaya Bay in the 1990s with the construction of a 7km- (4.5-mile-) long wall across the bay. It enclosed 3,000 hectares (7,400 acres) of tidal flats. By 2003 large areas of formerly flooded bay had been drained and turned into fields. The coastline has been transformed. Behind the wall, more tidal areas are slowly drying out ready for conversion. But not everyone is happy. Environmentalists say the old tidal flats were vital bird sanctuaries. And some farmers complain that they have lost valuable shallows where they once collected seaweed. There is now a campaign to tear down the new sea wall and restore the natural ecosystem. In early 2010, faced with mounting public pressure, the Japanese government seemed poised to halt further land reclamation, while opening the dyke's floodgates and allowing the seawater to return.

PRIPYAT Ukraine

1960s Pripyat was never a pretty city. A metropolis of some 50,000 people, it was built by the government of the Ukraine when that country was part of the Soviet Union to house workers at a power station 3km (2 miles) away. That power station was Chernobyl, which suffered the world's worst nuclear accident in 1986. It took the Soviet authorities some time to acknowledge the scale of the disaster. But four days after the accident, with fallout raining down on Europe – its cloud stretched as far as the hills of north Wales – the population of Pripyat was finally evacuated.

A deserted city

Pripyat has stayed evacuated ever since, along with a wide exclusion zone downwind of the plant. Radiation levels remain too high for people to return to this ghost city, which has become an unvisited museum of life in the old Soviet Union. It will probably be centuries before Pripyat is safe enough for people to return there without protective clothing. But, with humans virtually absent, nature has made a spectacular comeback. Trees grow in the radioactive soils, where children once played. Wild boar, deer, elk, beavers, and even wolves have invaded the city in their thousands. Unaware of the radioactivity that will doubtless shorten their lives, wild animals revel in the absence of humans. And behind it all the power station, with its core now in a concrete tomb, looms on the horizon.

DETROIT USA

1949 For much of the 20th century, Detroit was the bustling "motor city", among the United States' most populous, and the centre of both the American car-making industry and Tamla Motown music. Migrants flocked here. The mainly working class Hamtramck district, known locally as Poletown, became home to tens of thousands of East Europeans, many of them making Buicks, Oldsmobiles, and Cadillacs in the nearby General Motors plant. Its local church, St Cyril, with its school and Byzantine sanctuary, became a centre for the Polish community amid the packed streets.

Flight to the suburbs
But since the 1970s, the labour force at General Motors has much reduced, and the East Europeans have begun a second migration – from the packed streets of Hamtramck to the suburbs. The proportion of the local population who are Poles fell from 70 percent in 1970 to 20 percent by 2000. Their old homes have been demolished, leaving behind green spaces. St Cyril's Church held its last service in 1988 and was demolished shortly after this picture was taken in 2003. This abandoned urban landscape awaits a new use.

ST PETER'S BASILICA, ROME Italy

1996 Smog and other forms of air pollution have blackened and eaten away at the stonework and statues of most cities over the decades. The acidity of rain and fog, which can contain moisture as caustic as battery acid, dissolves marble and limestone. And soot from smoke and diesel blackens everything. This image shows a travertine marble statue of an angel on the 400-year-old façade of St Peter's Basilica in the Vatican, damaged by centuries of exposure to the smoke of Rome, topped up by the diesel fumes of the tourist buses that visit the Vatican.

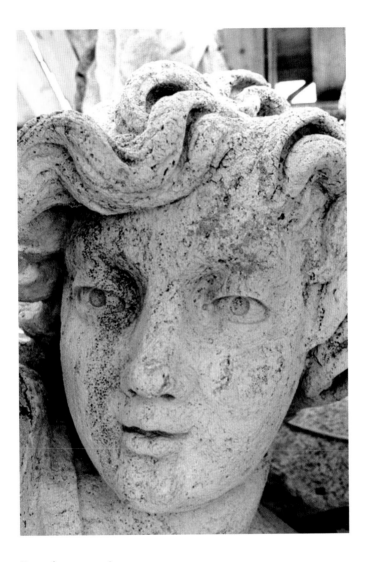

Papal restoration After 30 months of restoration the detail of the statue is greatly improved, and hidden tints of red and green that were added to the marble in the 18th century are once again on full display. Pope John Paul II unveiled the restoration of the façade in late 1999. The work was sponsored by a major Italian oil company – whose products would have helped to cause the damage. It had taken $5 million and almost three years of scrubbing and cleaning to remove the stains from the façade, which is regarded as one of the masterpieces of early Baroque architecture.

KUALA LUMPUR Malaysia

2005 As cities of the developing world go, Kuala Lumpur, the capital of Malaysia, has relatively clean air. Most days its gleaming new high-rise buildings, such as the Petronas Twin Towers seen here, are visible with ease from across the city of 1.5 million people. Malaysia, which plans to be a fully industrialized country by 2020, has attempted to get its first-world clean-air legislation in place ahead of time.

Forest fires

Legislation may be well meaning, but it can still be compromised. Kuala Lumpur is only 100km (60 miles) from the Indonesian island of Sumatra, which is clearing its rainforests faster than almost anywhere on Earth. The aim is to make room for palm oil plantations, and the favoured method is by setting fire to the trees. During the dry season – and particularly during El Niños, when the rainforests dry out – these fires often rage out of control. They create huge palls of smoke that are no respecters of international borders. On this occasion, noxious fumes from the forest fires filled Kuala Lumpur for several days, and the city's inhabitants could only hold their breath until the wind changed.

Clearing the skies When the Eyjafjallajokull volcano in Iceland began erupting in April 2010, few imagined the consequences. A cloud of volcanic ash was blown into the upper atmosphere where it began drifting south and east into some of the busiest aircraft corridors in the world. Air traffic across most of western Europe, as far south as Italy, was shut down for almost a week, because of fears that the ash would shut down jet engines. The disruption left hundreds of thousands of would-be passengers stranded across the world and airlines $1 billion out of pocket.

Forces of Nature

Whatever mankind's impacts on our environment, the forces of nature are often of far greater ferocity. The shifting plates that make up the Earth's crust constantly cause earthquakes and unleash molten rock and ash from the planet's core through volcanoes. Sometimes those emissions can be enough to shroud the Earth in particles that cool the atmosphere for months or even years – as Mount Pinatubo in the Philippines did in 1991.

In the distant past, such events have been very bad news indeed for humans. When Mount Toba in Indonesia blew its top 73,000 years ago, the cold was so intense for so long that *Homo sapiens* almost died out. Most of the time, however, the danger to humans from these geological forces comes more immediately – from the destructive power of exploding mountains, shaking rock, flowing lava, and sometimes from the avalanches, tidal waves, and collapse of buildings that they unleash.

Volcanoes are a constant threat. In 1980 Mount St Helens in the USA lost 400m (1,300ft) from its summit during an eruption, creating the largest landslide in recorded history. Most of the Caribbean island of Montserrat has been out of bounds since its volcano blew in 1997, and the majority of its former inhabitants have never returned.

Earthquakes are, if anything, even less predictable. Usually they kill when shaking buildings collapse, as in the disaster in Haiti in early 2010, when at least 90,000 and perhaps as many as 230,000 died in the rubble. But in mountains, quakes can cause lethal landslides and avalanches. Most of the dead from a quake in the Pakistani foothills of the Himalayas in 2005 perished when rocks rained down on them. And earthquakes on the sea bed can cause tidal waves that roar inland sweeping all before them. On Boxing Day 2004, a tsunami following a quake beneath the Indian Ocean drowned more than a quarter-million people.

Floods, earthquakes, and volcanoes are killing more people each year. This is largely because our ever more crowded planet puts more and more people in harm's way. Most people live on coasts, vulnerable to floods, for instance. Some even live below sea level, trusting, like the people of New Orleans, to the protection of sea defences. Volcanic regions often have extremely fertile soils, enriched over centuries by lava flows, and they attract people. An added painful twist is that it is the poor in many cities that so often end up living on land vulnerable to floods and landslides. Even in known earthquake zones, few buildings are designed and built to resist shaking.

But there is a growing fear, too, that humans can sometimes trigger "natural" disasters. By deforesting hillsides and draining marshes, we are artificially increasing the intensity of river floods. Global warming may be increasing the risk of avalanches from melting ice and thawing permafrost in high mountains. Is that what happened on Mount Cook in New Zealand in 1991? Climate change may also be increasing the intensity of hurricanes – this is a hot topic after Hurricane Katrina broke levees and unleashed unprecedented floods in New Orleans in 2005. When nature conspires with the hand of man to unleash such forces, we should take great care.

SURTSEY ISLAND Atlantic Ocean

1963 Early one November morning in 1963, a crack opened in the Earth's crust on the bottom of the Atlantic Ocean, about 30km (19 miles) south of Iceland. Red-hot magma burst upward into the ice-cold water, while water fizzed down into the crack. Fishermen began to notice black smoke and then a stench of sulphur. For months the eruptions continued on the sea bed. The sea boiled, ash was blown into the air, and lightning shot between clouds of volcanic ash, until eventually a new island began to solidify above the waves. The island of Surtsey, named after a mythical Norse fire-eating giant, was born.

New growth
Slowly the eruptions ceased and the island began to cool. During Surtsey's first spring, scientists stepped ashore to find that wildlife had got there first – a lone fly and some seeds dropped by passing birds. The following year they found white flowers of sea rocket, which had sunk its roots into the ash. Lichens and mosses, insects, and even sea-bird colonies have followed. It is, some say, like the world being born anew. But for how long? When the eruptions stopped, Surtsey had a surface area of 2.5sq km (1sq mile), but the Atlantic swell is already eating away at its cliffs and beaches. The island has halved in size since 1967. Within a century or two it may disappear beneath the waves once more.

HEIMAEY ISLAND Atlantic Ocean

c.1970 Heimaey Island is just off the south coast of Iceland. Its only settlement, and one of Iceland's biggest fishing ports, is Vestmannaeyjar, which sits in the shadow of a volcano called Eldfell. The volcano was once thought to be extinct, but as events at nearby Surtsey showed, the area south of Iceland is still volcanically very active.

Race against time

Early one January morning in 1973, Eldfell began to spit showers of red-hot ash into the cold air and onto the outskirts of the town. By dawn, the island's population of 5,000 had fled to the mainland. Meanwhile, the eruption continued. Some 65 buildings disappeared beneath the black cinders, and another 300 burned to the ground. But, worse still, a slow tongue of lava continued to flow out of the volcano, threatening to block the harbour and destroy the town's economy. For four months volunteers worked in relays to spray cold seawater onto the hot lava to solidify it and slow the advance. Just in time, the volcano quietened. The harbour was saved and the fishing community could start to rebuild. And the millions of tonnes of ash they cleared from the town came in handy to make a second runway for the island's tiny airport.

MOUNT ST HELENS USA

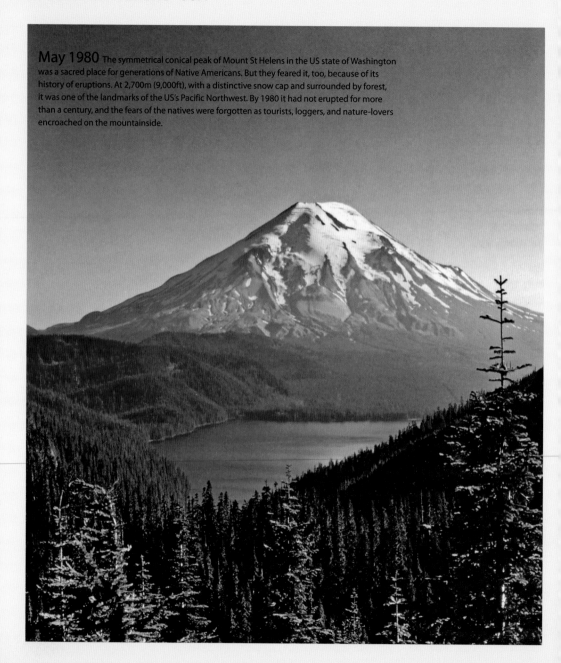

May 1980 The symmetrical conical peak of Mount St Helens in the US state of Washington was a sacred place for generations of Native Americans. But they feared it, too, because of its history of eruptions. At 2,700m (9,000ft), with a distinctive snow cap and surrounded by forest, it was one of the landmarks of the US's Pacific Northwest. By 1980 it had not erupted for more than a century, and the fears of the natives were forgotten as tourists, loggers, and nature-lovers encroached on the mountainside.

Massive eruption
At 8.32am on 18 May 1980, after weeks of rumbles and emissions of ash, Mount St Helens erupted with huge force. The volcanologist David Johnston, who screamed into his radio "This is it!" as the mountain burst, was dead two seconds later, obliterated by an advancing wall of rock, ice, and trees. The eruption had been triggered by an earthquake beneath the mountain. The mountain's north face collapsed in an instant, creating the largest landslide in recorded history. It buried everything in its path to a depth of 200m (650ft) and more. The blast wave snapped fully grown trees 20km (12 miles) away; 57 people died. When the clouds of ash cleared, shocked onlookers saw that the snow-covered cone had become a flat-topped snowless stub with a peak 400m (1,300ft) lower than before. Two years later the US Congress created the National Volcanic Monument, a vast national park around the mountain, where nature is being left to recover as best it can.

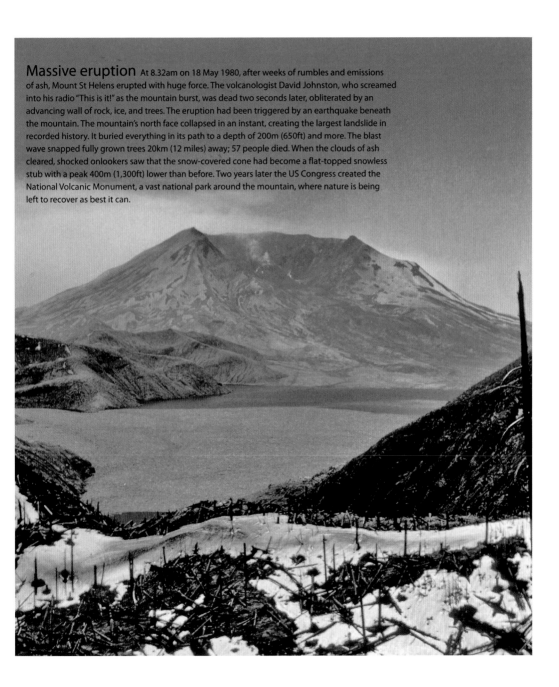

MOUNT ST HELENS USA

1973 This infra-red satellite image shows Mount St Helens before
the eruption: a white summit of snow and ice surrounded by muddy
slopes and a wider mountain region of forest, which shows up green.

Fallen debris
Ten years later, this new image shows the rock, mud, ash, and trees from the explosion in shades of grey. They are smeared across mountains and valleys to the north of the volcano for more than 10km (6 miles). A series of lakes, shown black, has formed where rivers were blocked by debris, in places 170m (560ft) deep.

MOUNT KILAUEA Hawaii

1983 Most volcanoes are brooding presences. There may be hundreds or even thousands of years between each explosive eruption. That is enough time for many of the obvious indicators of volcanic activity to disappear, though they are always liable to erupt again. Others return to life rather more frequently, though usually less violently. One such is Mount Kilauea on the southernmost island of Hawaii. It is one of the world's most permanently active volcanoes. This is Puu Oo, its easternmost vent, which began spewing hot lava down the mountainside in 1983, and has been erupting slowly ever since.

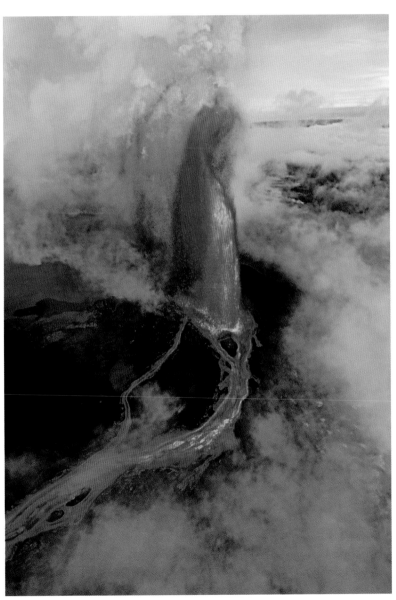

Constant lava flow

Here is Puu Oo again in 2001. So far, over more than two decades, the current eruption has extruded some 2cu km (0.5cu mile) of lava from beneath the earth and covered more than 100sq km (40sq miles) of Kilauea's southern slopes. In the process it has extended the island's shoreline and destroyed 200 houses, a visitors' centre, a 700-year-old Hawaiian temple, and 13km (8 miles) of roads. There are no signs of it stopping any time soon. This persistent nonexplosive activity – and the fact that it creates shallow slopes rather than the steep slopes of a conventional volcano – makes Kilauea a favourite among volcanologists. It is known as the "drive up" volcano, because even its central crater can be reached by car.

MONTSERRAT Caribbean

1993 Montserrat is one of the smallest of the leeward islands in the Caribbean, just 16km (10 miles) long. It is also one of the last surviving specks of the British Empire, ruled from London. In the early 1990s it had a population of 12,000 people, mostly living in Plymouth, the sleepy port capital on its southern shore. The island had a tourist industry and its land was fertile, in large part because of the volcanic soils provided by the island's largest geographical feature, the Soufrière volcano behind Plymouth. But in 1997 the volcano erupted for the first time in recorded history. A cloud of ash, pictured here in the first minutes of the eruption, rose into the air behind Plymouth…

Blanket of ash Within hours of the eruption, huge volumes of ash and boulders rained down on Plymouth. This picture was taken later the same day, as residents hurriedly left from the town jetty. By the time the eruption was over the whole southern half of the island was blanketed, and parts of the town were buried in ash more than 3m (10ft) deep. Plymouth's airport was destroyed. Today just 4,000 people remain on the island, all living in the north, which was previously largely uninhabited. The southern half is still out of bounds and regarded as permanently unsafe because of sporadic further eruptions.

MAHABALIPURAM India

2004 The temple of Mahabalipuram, on the seashore of southern India, is famous. It is a World Heritage Site, with some of the earliest-known examples of Dravidian architecture dating back to the seventh century AD. But locals always said there were six more large temples that still lay buried beneath the sands. The British traveller John Goldingham, who visited the south Indian coastal town in 1798, when it was known to sailors as the Seven Pagodas, documented the stories. Locals told him that Mahabalipuram was once a large city so beautiful the gods became jealous and sent a flood that swallowed it up in a single day.

Asian tsunami

Then came the tsunami of 26 December 2004. It surged across the sands, and when it retreated, it revealed the fabled remains of temples and an ancient port. Archaeologists rushed to the site. Within weeks of the tsunami, they were at work charting the newly exposed remains. A half-completed rock relief of an elephant is now drawing large crowds. Many more remains lie offshore, with the silt that once hid them removed by the tsunami's surge. Archaeologists now believe that it was another tsunami, in the 13th century, that first covered up the buildings.

MOUNT COOK New Zealand

1991 Mount Cook is the tallest peak in New Zealand. Its sharp summit is also known by its traditional name of Aoraki, or cloud-piercer. The mountain is wrapped on all sides by two large glaciers that make it an attractive challenge for mountaineers, but also a hazardous one. The steep slopes are made unstable by the ice hanging from them. It could fall off at any time, perhaps taking rock with it.

Mountain-top avalanche

On 14 December 1991 that is what happened. That night mountaineers waiting in huts for a morning ascent heard great rumbles from the mountain and saw sparks as rock clashed against rock in the dark. Something had given way on the summit and by daylight 14 million cu m (500 million cu ft) of rock, ice, and snow had tumbled down from its east face. Some of the rock ended up 7km (4.5 miles) away on the Tasman Glacier below the mountain. The avalanche dramatically changed the shape of the peak and reduced its height by about 10m (33ft), to 3,754m (12,316ft). Further avalanches are expected, and many believe that global warming is likely to destabilize many more ice-covered mountain summits around the world.

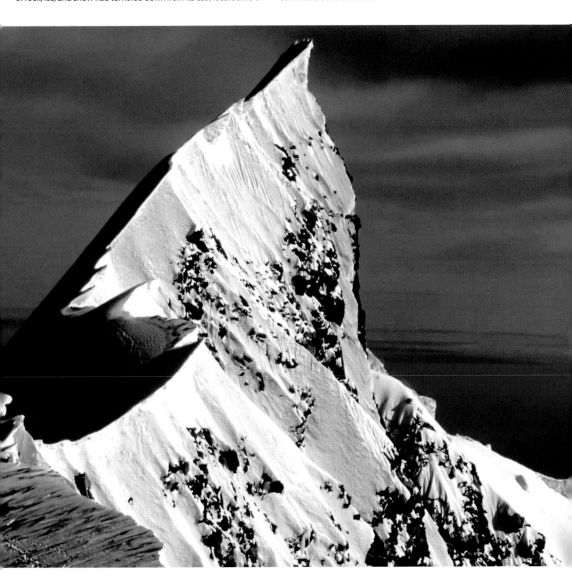

PORT-AU-PRINCE Haiti

2004 Haiti, the poorest nation in the Americas, was an economic and ecological mess before the January 2010 earthquake hit the Caribbean state. But at least its capital, Port-au-Prince, had an impressive National Palace, official home of the president, most recently René Préval. The palace was designed by one of Haiti's most distinguished architects, Georges Baussan, and was similar to many grand renaissance-style buildings in France, the country's former colonial occupier. Construction of the palace began in 1914, but the half-completed structure burned down the following year during riots that ousted the president. Finally completed in 1920, the building was twice the size of the White House, home of the US president.

Shaken to pieces Not much survived the earthquake that struck Haiti on the afternoon of 12 January 2010. Its epicentre was just 16 km (10 miles) from Port-au-Prince. The second floor of the palace entirely caved in and the cupola collapsed. Nobody died here, however. The president was at his private home and survived. Estimates of how many Haitians died that day range from the government figure of 230,000 to independent reports of just over 90,000. In addition, a quarter of a million were injured and up to a million left homeless, creating a massive humanitarian crisis that had aid agencies floundering. France has offered to rebuild the palace.

ANCHORAGE, ALASKA USA

1959 Anchorage began as a base for construction of Alaska's first railway. It was built where the railway met the sea, with the first streets laid out in 1914. For a long time it had no agreed name. Eventually, the US Postal Service gave up on locals and declared it to be Anchorage. Urban expansion began seriously in the 1940s with the construction of an air force base nearby. The town gained further importance in 1959, the year this picture was taken on Fourth Avenue, when Alaska was incorporated as a full-scale state within the Union.

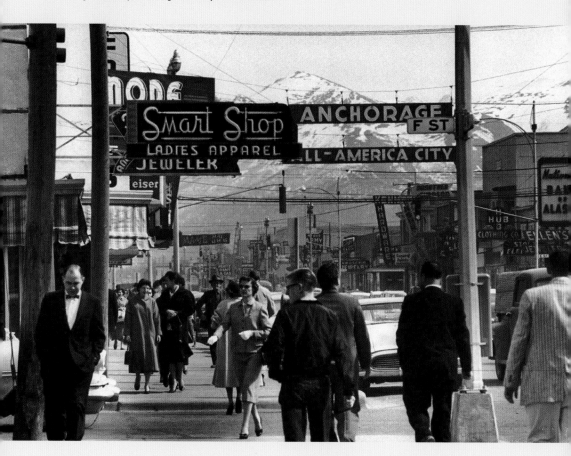

After the quake

On 27 March 1964, Good Friday, the five-year-old state of Alaska was hit by one of the biggest earthquakes of the 20th century. It registered a magnitude of 9.2. The change to the landscape from this huge shift in the Earth's plates was immense. This photo shows Fourth Avenue two days after the quake: the road has collapsed so completely that cars have ended up some 6m (20ft) below the former street level. Yet, amazingly, while one side of the avenue was wrecked, the buildings on the other side are largely intact. Altogether 131 people died. High-rise construction is still banned here because of the quake threat.

MARYSVILLE Australia

2009 "Black Saturday" saw Australia's worst ever bushfires and worst peace-time disaster. On 7 February 2009, more than 400 fires were recorded across 4,500sq km (1,750sq miles) of New South Wales. They resulted in 173 confirmed deaths, injured more than 400 people, and destroyed more than 2,000 homes. The conditions could not have been worse. The forests were a tinderbox after almost a decade of drought, temperatures were in the mid-40s°C (110–120°F), humidity at a record low, and wind speeds were up to 100km (60 miles) per hour. When the winds felled power lines and ignited the vegetation, the result was a perfect firestorm. The small town of Marysville, northeast of the state capital Melbourne, was almost completely destroyed and 34 of its 500-strong population died in the flames.

One year on Life goes on. The recovery is under way.
New tree growth is sprouting across the hills of New South Wales
on the first anniversary of "Black Saturday". And people have mostly
returned too. In Marysville, they have rebuilt many of the shops. But
memories of the fire will be harder to erase, and the cemetery has
a number of new occupants.

MAKHRI Pakistan

2002 The mountain villages of Pakistani Kashmir, in the foothills of the Himalayas, are among the most remote areas on Earth. The villages huddle in steep valleys and it can take many days to reach them from the outside. The people are extremely poor. They have few roads, schools, or health clinics and only sporadic contact with any form of government. Here, photographed by satellite in September 2002, is the village of Makhri. It sits on a rare area of flat, fertile land beside the River Neelum, a tributary of the River Indus, and faces a tall mountain ridge on the opposite bank.

Crippling landslide
In October 2005, an earthquake shook these mountains. Besides levelling buildings, it caused hundreds of landslides on the steep mountainsides that engulfed entire villages and blocked the roads that relief workers attempted to take over the following days. One of the worst slips saw the collapse of the western face of the mountain opposite Makhri. The landslide, visible here as mounds of grey rock and mud, blocked the large meander in the River Neelum, forcing it to take a new course. The formerly blue river is now chocolate brown with the accumulated debris of this and other avalanches. Some 200 people died here.

BRITISH ISLES

2009 The British Isles are not always cloudless. But when they are, they usually look like this: green vegetation, purple heather, a smidgeon of dry and barren brown during droughts, and a few patches of snow and ice on high ground in winter – all surrounded by the blue sea.

Winter's grip

But January 2010 was different. Global warming seemed to be put on hold by the coldest winter since 1963. The entire British Isles was blanketed by snow for several days. A few patches of white here in this satellite image, taken on 7 January at around noon, are clouds. But mostly it is snow.

TWELVE APOSTLES Australia

2005 These rugged limestone stacks, known as the Twelve Apostles, stand in the sea within the Port Campbell National Park, 220km (135 miles) southwest of Melbourne in Australia. They are among Victoria state's top tourist attractions, easily visible from the spectacular Great Ocean Road. They stand as high as the cliff tops close by, an apparent symbol of the enduring ability of rocks to withstand the constant battering of ocean waves. But in fact their name has long been out of date. There were only nine stacks remaining when this photograph was taken…

Instant erosion …And less than one minute later, when this second photograph was taken, there were only eight stacks. The 50m- (165ft-) high stack in the foreground on the left had collapsed into the sea before stunned onlookers – a testament this time to the power of the ocean. Will the apostles eventually all disappear?

It might seem so. But the land behind them is also being eroded, and as the sea gnaws away at fractures and other weak spots in the cliffs, leaving behind the hardest rock, new stacks could be created at any time. In a few decades, say geologists, there could be 12 apostles again.

ACEH Sumatra, Indonesia

2003 People have lived along the coast of Sumatra for thousands of years. They have long since cleared most of the forests and mangroves to make way for fields and rice paddy. But Aceh Besar, the northernmost district of Aceh, the northernmost province on the island, nonetheless remains largely rural. This patch of lowland, photographed in early 2003, is a lush green backwater in every sense. The rice paddy shows in light green. The water on the bottom left of the picture is a small freshwater wetland about 10km (6 miles) inland from the shores of the Indian Ocean.

Tidal wave On Boxing Day 2004, a tsunami struck the Indian Ocean. An underwater earthquake, the largest anywhere in the world for 40 years, caused devastation right around the Indian Ocean, but nowhere suffered more than Aceh, where an estimated 130,000 of the tsunami's quarter-million victims died. Within minutes the landscape was dramatically transformed. Here, the sea swept across the coastal town of lho-nga and inland, engulfing the small wetland and consuming all the low-lying paddy and many settlements beyond. A few homes and roads were left marooned on higher ground.

BANDA ACEH Sumatra, Indonesia

2004 On Boxing Day 2004, the world ended with fewer human inhabitants than it began the day – a first since the start of the 20th-century population boom. More than two-thirds of the deaths from the tsunami occurred in Aceh, many here in the provincial capital of Banda Aceh. Four days after the tsunami struck, the streets of Banda Aceh are full of debris washed inland by the tidal wave. Many of the city's quarter of a million people lay dead.

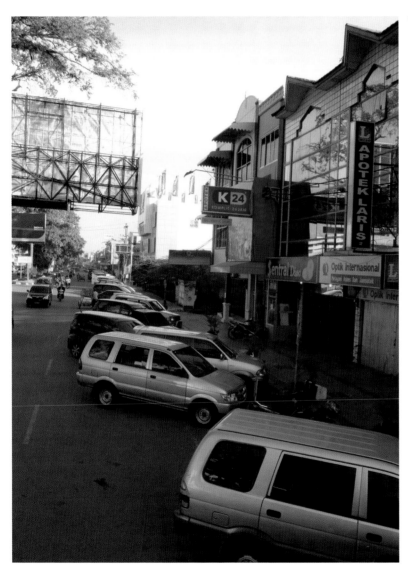

The "New Bali"

Five years later, and the same street is cleared, as if nothing had happened. Reconstruction has been so rapid and effective that the town of 260,000 people is now a tourist destination. Visitors come for diving and to sunbathe on beaches that were killing zones in 2004. Many join the "tsunami tours" that visit ships stranded sometimes several miles inland. Such is its popularity that some call Banda Aceh "the new Bali".

NEW ORLEANS USA

2005 The suburbs of New Orleans suffered as much as the city centre when Hurricane Katrina swept through southeastern Louisiana. This is the damage fully a week after the levees on the Industrial Canal, a major artery between the Mississippi and one of the delta's main lakes, were breached in three places. The neighbourhood is near the boundary between the city and St Bernard parish, just east of the downtown area. Only the roofs of houses are visible. Anyone who did not manage to get out or scramble onto a roof is dead.

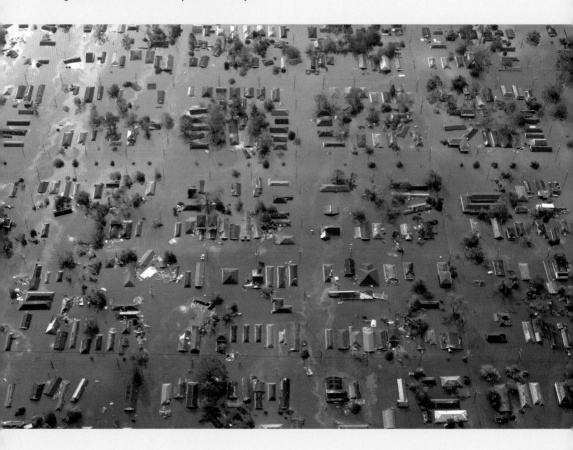

Flood aftermath This is the same area, 12 days later. The
waters are retreating. Some of the roads are dry, though others are
still flooded. What remains of the houses is emerging, waterlogged
and full of mud. A few people return to collect any belongings that
survived. But this is a rare example of almost an entire major city –
more than 80 percent by most counts – being inundated.

NEW ORLEANS USA

2005 A school bus sits on a flooded street in the Ninth Ward of New Orleans on 12 September 2005. It is two weeks since Hurricane Katrina ripped through the city on the US coastline of the Gulf of Mexico, destroying its flood defences. Clearly visible is the breach in the concrete levee that once held back the water in the Industrial Canal, which links the Mississippi River to one of the major lakes on its delta. In similar scenes, some 1,800 people died across a city built largely below sea level.

Revival Ten months later, the levee on the Industrial Canal has finally been repaired and the water held back. Shipping is back on the canal, the power lines are functioning again, and New Orleans residents are struggling to rebuild their city. But the cost of repairing the damage is put at over $100 billion. Since 2005 there have been calls to block the canal with a dam to prevent storm surges ever again rushing through the city.

BETSIBOKA RIVER Madagascar

2003 The Betsiboka is the largest river in Madagascar. It drains a swathe of the island's mountainous north, an area once covered by rainforests containing wildlife unique to the island, such as its famous lemurs. This picture, taken from the international space station in September 2003, shows the river in the dry season. It is draining through a large delta into the ocean, which can be seen on the top left. Upstream, largely out of shot, the rainforests have mostly been chopped down.

Cyclonic effect

The second picture shows the same scene six months later, a fortnight after tropical Cyclone Gafilo swept across Madagascar bringing heavy rains. The denuded hillsides are now hugely vulnerable to erosion after rain. And the evidence is here. Rapid runoff from the hillsides has flooded the river and brought down millions of tonnes of soil. The soil has clogged the delta, stained the river bright orange, and sent a vast plume of sediment into the estuary. It is as if the island is bleeding into the ocean. The constant arrival of silt down the river is clogging up the estuary so that ocean-going ships that could once travel upstream must now berth on the coast.

SOUTH DAKOTA USA

1935 The "dust bowl" took a terrible toll across the American Midwest in the 1930s. When drought hit the dry plains, crops failed, fields turned to deserts, the skies filled with clouds of dust, and millions of destitute farmers headed west to become labourers in factories and plantations in California. John Steinbeck's epic tale *The Grapes of Wrath* described the disaster best. Here, at the height of the drought, sand dunes up to 2m (7ft) deep have formed across formerly ploughed fields and encroach on a farmstead in Beadle County, South Dakota.

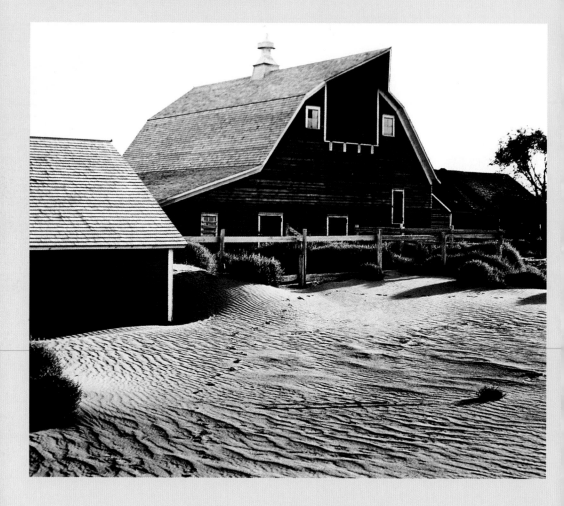

Soil conservation
Many believed that the "desertification" of the Midwest would be permanent. But the federal government set to work with a soil conservation programme, part of the New Deal for America's poor. It achieved extraordinary results. Here is the same farmstead two years later. An emergency cover crop of Sudan grass has captured moisture and stopped the soil blowing away. Later, boreholes were sunk to abstract copious amounts of underground water from beneath the High Plains. Now farmers can irrigate their crops during dry spells. There's just one problem: the aquifers are emptying, so the region may again be living on borrowed time.

ATACAMA DESERT Chile

Early 1990s The Atacama desert in northern Chile is the driest place on Earth. Average rainfall is just 1mm a year, but many places go for many years – in one recorded case for 40 years – without a drop of rain. The desert stretches for 1,000km (600 miles) parallel to the Pacific coast, in a rain-shadow between coastal mountains and the high Andes. The Atacama's salt plains and lava flows are about the most inhospitable place for life on Earth. Recent research suggests the soils are so devoid of microbes that they are more like those on Mars than any on Earth.

Desert blooms Just occasionally it does rain in parts of
the Atacama, and the results can be extraordinary. Even if the rains
come only once a decade, seeds that have remained dormant in
the ground for years can germinate within hours and burst forth in
amazing carpets of colour. Species often flower in sequence, so that
sometimes a valley will be purple one day, yellow the next, and blue
after that. Usually these desert blooms appear with the occasional
heavy rains caused by El Niño, a periodic reversal of normal climatic
patterns in the Pacific.

In harm's way US Marines patrol through Now Zad in Helmand province, Afghanistan, at the height of action to curb the rise of the Taliban insurgency in mid-2009. Taliban fighters have booby-trapped the landscape, burying explosive devices through the town since citizens abandoned it after earlier fighting. So the marines patrol in single file behind a mine sweeper to avoid stepping on the hidden explosives.

War and Conflict

Arguably, wars have disfigured the modern world even more than the destruction of the natural environment. The 20th century may one day be known more for the extraordinary carnage of its wars than for anything else. This chapter looks at the physical impact of the industrial-scale annihilation of people – but also at subsequent efforts at reconstruction and commemoration.

Look at the remains (or rather the absence of remains) of a Belgian village called Passchendaele, where three-quarters of a million soldiers died over five months during World War I in an ultimately meaningless contest for the village. Or try to imagine the human tragedy behind the simple street scene in Leningrad, where half a million died of starvation and disease during a 900-day siege during World War II.

Wars mostly kill people, but sometimes the environment is directly attacked for strategic advantage. In this chapter we show two examples of the latter. One is the extensive damage done to the rainforests of southeast Asia by the relentless US bombardment with the herbicide Agent Orange. Its purpose was to expose the jungle supply routes of the Vietcong during the Vietnam War. Another is the deliberate draining of the Mesopotamian Marshes – the largest and most fecund wetland in the Middle East – by Saddam Hussein during the 1990s. His purpose was to force out the rebellious Marsh Arabs, who had occupied the marshes for thousands of years.

Nuclear weapons are indiscriminate destroyers of the environment and humanity. The obelisk in the New Mexico desert doesn't look much, but it was here that the first atomic bomb was tested in July 1945. The success of that test not only stunned the creators of the bomb, it also ensured the destruction of Hiroshima and Nagasaki in Japan – and the loss of 200,000 civilian lives – less than a month later.

But maybe this chapter can offer some hope. For it shows the power of the human spirit to heal wounds. Looking at the pictures of Caen and Dresden today, it is hard to imagine that either city was horrifically firebombed little more than half a century ago. Even Hiroshima has recovered, although its oldest citizens continue to die from diseases caused by the radiation released the day that "Little Boy" was detonated over the city.

The barricade at the Brandenburg Gate in Berlin was a symbol of the Cold War, and its dismantling has become a symbol not just of the reunification of Berlin and of Germany, but of the reunification of Europe and the end of the most dangerous stand-off in human history. Not that Europe has been entirely at peace since. It has been scarred by a decade of war in the Balkans following the collapse of Yugoslavia – a complex series of conflicts symbolized here by the death and rebirth of the Mostar Bridge in Bosnia.

War intensifies in Afghanistan. Sabres rattle over Iran. And the Middle East continues to fester. Many believe that, while the conflict between Israelis and Palestinians persists, the fuel exists for a wider conflagration between the Islamic world and the West. And in that dispute nobody is safe – as workers at the Twin Towers in New York discovered one bright sunny morning in September 2001.

PASSCHENDAELE Belgium

1917 Few place names chill the bones of historians more than Passchendaele. It has become a synonym for the horrors of war. In 1917, at the height of World War I, British, Canadian, and Australian soldiers wanted to cut through German lines in Belgium to capture a German submarine base on the coast. They chose to do it at Passchendaele, a small Belgian village in a marsh near Ypres.

Obliteration
The battle lasted five months. Tanks, mines, and artillery were deployed on a scale never seen before. So was mustard gas. The constant bombardment turned the marsh into a huge quagmire. Many soldiers drowned in the mud after falling from duckboards. The village was captured, and this is what was left. Nothing. The destruction was as total as if an atomic bomb had struck the village. Aerial photographs such as this showed a million shell holes within just more than 2sq km (0.8sq mile). That is one for every 2sq m (22sq ft) of sodden ground. The Germans lost 270,000 men, and the British and their allies not far short of half a million. Some 90,000 bodies were never identified, and 42,000 never recovered. The war poet Siegfried Sassoon wrote: "I died in Hell; they called it Passchendaele."

THE SOMME France

1916 It was the first day of the Battle of the Somme, one of the most bloody conflicts of World War I. At 7.30am on 1 July the soldiers of the Newfoundland regiment, who had just arrived from Canada, left their muddy trenches dug into French soil at Beaumont-Hamel and advanced on German lines. Officers told the young recruits that the Germans had been decimated by five days of artillery bombardment. Instead, their opponents had survived the onslaught relatively unscathed, and cut them down with machine-gun fire. Of the 800 Canadians who went "over the top" that day, 310 died and only 110 returned unhurt – just one skirmish on a day when 20,000 died along the 45km (28-mile) front. Altogether, the Battle of the Somme lasted five months and when a halt was finally called, more than a million had died.

Remembrance

Today Beaumont-Hamel is once again a quiet hamlet, 9km (5.5 miles) north of the town of Albert in northern France. But on its outskirts, the trenches and artillery-blasted fields are a reminder of what happened there. And the 30 hectares (75 acres) where the battle was fiercest is one of the most visited of all the memorial grounds for the dead of the Somme. The Beaumont-Hamel Newfoundland Memorial Park, opened in 1925, contains three cemeteries with 630 graves. And beneath the grass lie the unmarked remains of the 90 men who fell that day but whose bodies were never recovered.

LONDON England

1940 Cannon Street, in the heart of the City of London, England, is known today mainly as one of London's railway termini. But on 29 December 1940 it was firebombed by German aircraft at the height of their Blitz on the capital of the British Empire. After that night, and a further bombardment four months later, the street was reduced to rubble, as this view here from the top of St Paul's Cathedral, which miraculously survived, shows. In the distance, another survivor was Tower Bridge, over the River Thames.

In business Here is the same view today, with Cannon Street rebuilt and Tower Bridge still visible in the distance. Sixty years on the British Empire may have gone but the City of London stands beside New York as one of the two primary financial capitals of the world. Unusually for a major central business district, most of the new buildings are on a similar scale to those they replaced. What is different is their colour. The wartime buildings were blackened by decades of coal smoke that made London one of the smog capitals of the world. Vehicle fumes still fill the streets but the black coal smuts are gone and the new buildings stay the colour of their stone.

CAEN, NORMANDY France

1944 The Battle of Caen, between advancing Canadian forces and Germans in slow retreat, was one of the centrepieces of the final phase of World War II. It began days after the Canadians came ashore during the D-Day landings on the Normandy beaches, and lasted for two months. The city's medieval heart was burned out when a rain of bombs on 6 June 1944 set the city burning for 11 days. As the Germans continued their resistance, most of the citizens took refuge in a few buildings, including the churches, that stood like beacons amid the rubble.

Reconstruction The heart of Caen has been almost completely rebuilt since the war. The churches and the ancient castle ramparts are almost all that remains of the old city. But some of the original streets live on as modern shopping precincts, cobbled and pedestrianized. The bunting in this picture celebrates the city's survival, and displays the flags of the liberating forces. Old soldiers and tourists alike use the hotels as a base to visit the famous landing beaches along the shore to the north of the town.

HOHENZOLLERN BRIDGE, COLOGNE Germany

1942 Intense Allied bombing in 1942 destroyed much of Cologne, Germany's fourth-largest city, causing severe damage to the famous Hohenzollern Bridge across the River Rhine. Behind it, the twin-spired cathedral, one of the largest Gothic buildings ever constructed, was almost the only building in the city's heart to survive. The bridge was swiftly repaired. But in March 1945, with the Allied forces sweeping east, the German forces abandoned Cologne and blew up the bridge as they went. In a last, desperate effort to halt the Allied advance, they destroyed most of the other bridges across the Rhine between Bonn and the Netherlands as well. But, in the process, they condemned to capture some 50,000 German troops stuck on the west side of the river.

The new bridge The bridge has been rebuilt once more. While earlier versions carried road as well as rail traffic, the post-war version is reserved for trains and pedestrians. But it remains a central link between the city's east and west sides, and a main route from Cologne to its hinterland on the western side of the river.

FRAUENKIRCHE, DRESDEN Germany

1952 The Allied firebombing of Dresden back in 1945 was some of the most intense in World War II. Much of the southeast German city was flattened – including the totemic Dresdner Frauenkirche, a baroque 18th-century Lutheran church in the heart of the city. The church survived 48 hours of bombing, but finally succumbed to temperatures of up to 1,000°C (1,800°F) created by the fires that raged through the city. Its architectural crowning glory, a 12,000-tonne sandstone dome held up by eight pillars, gave way when the pillars glowed bright red and exploded.

Reconciliation In communist East Germany the ruins were left untouched as a war memorial until the early 1980s, when the site became the focus for a civil rights movement protesting against the government. It became one of the sparks for the reunification of Germany. After reunification, an international movement developed to reconstruct the church, as an act of reconciliation between once-warring nations. Reconstruction began in 1993 using the original plans and some 8,000 of the original, charred stones, which are clearly visible here. The exterior was completed in 2004, and the church reconsecrated the following year. Once a month an Anglican Eucharist is held.

LENINGRAD Russia

1942 The Siege of Leningrad was one of the worst horrors on the Eastern Front during World War II. This image says much about how the extraordinary can become normal in times of war. A woman pulls a shrouded corpse on a sledge down snow-covered Nevski Prospekt, the city's main street – the event goes barely acknowledged by passers-by. The siege lasted for 900 days. The first winter, when this picture was taken, was one of the coldest of the century. An estimated quarter of a million citizens died of cold and starvation. And, altogether, half a million may have died before the Germans gave up and went home in early 1944.

Metamorphosis
Much has changed. The Germans have long gone, but so, too, has Communism. The city has reverted to its pre-Soviet name of St Petersburg and this is the new-look Nevski Prospekt. The huddled masses are replaced by Japanese cars and Western-style advertising on what the city now hails as "Russia's most famous street", the home of Tchaikovsky and Nijinsky. What has stayed the same is the street's buildings, most of which date back long before Lenin, to the days of the czars. The winters are mostly warmer now, too.

TARAWA Pacific Ocean

1943 Tarawa is a clutch of 24 tiny coral islands huddled around a sleepy lagoon right on the equator in the central Pacific – an atoll that became a battleground in World War II. Here is the scene in November 1943 as US forces fought to retake the islands from some 5,000 Japanese, who were heavily dug in. The marines secured the atoll, but only after three days of fierce fighting that left the beaches covered in dozens of bodies and huge amounts of military hardware.

Battle scars
Tarawa was, until 1979, the capital of the British-owned Gilbert and Ellice Islands, although London rules counted for little in wartime. Now it is the capital of the independent Republic of Kiribati, whose main source of revenue is tourists and revenues from selling rights to fish in the vast swathe of the Pacific that it controls.

But, however much its politics have changed, many of the remnants of its brief moment of infamy still litter the beaches. Far from detracting from its value as a tropical paradise, the remains of the Battle of Tarawa and the Japanese defensive installations are today one of Tarawa's principal tourist attractions.

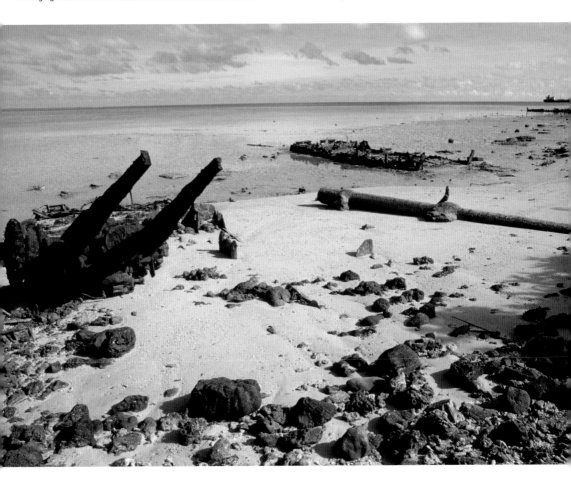

TRINITY SITE, NEW MEXICO USA

1945 Before the USA dropped its atomic bombs on Hiroshima and Nagasaki in Japan, at the end of World War II, it tested a prototype device in the US desert. Here is the site of the world's first atomic bomb test. Called the Trinity Site, it is 300km (186 miles) south of Los Alamos in New Mexico. The tracks lead to where the bomb, hanging from a 50m (165ft) tower, was detonated at dawn on 16 July 1945. It exploded with the force of 19,000 tonnes of TNT, instantly vaporizing the tower and creating a crater, which appears small in this picture but is 300m (985ft) across and 3m (10ft) deep. The resulting mushroom cloud reached 12km (7.5 miles) into the stratosphere. Nothing survived within 1km (0.6 mile) of the blast. Observers 9km (5.5 miles) away were knocked to the ground by the blast wave. The sand around the crater was melted to a green radioactive glass that shows dark in the area surrounding the crater. They called this new substance Trinitite.

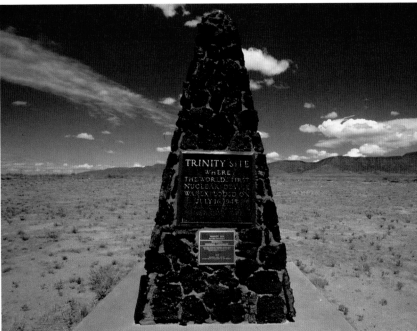

Monument In 1952 military engineers bulldozed the site of the explosion and buried the Trinitite, although collectors removed some samples first. Ground Zero is marked today by an obelisk that visitors are allowed to see twice a year. Radiation levels remain ten times the normal level. This is the site where the atomic age began and the world saw the first mushroom cloud.

BAGHDAD Iraq

2003 It was one of the iconic moments of the war in Iraq.
On 9 April 2003, before the world's cameras, local citizens and the
US military together toppled the 12m (40ft) statue of Saddam
Hussein. The moment symbolized the end of his 24-year rule.
In the background is the 14th of Ramadan Mosque in Firdos
Square in the heart of Baghdad.

Freedom statue

A month later, in Saddam's place on the plinth in Firdos Square, the "freedom statue" was erected. This sculpture, paint-spattered by the time this picture was taken two years later, was the idea of an artist's group called Najeen, which formed as an underground movement during the days of Saddam. Its name means "survivor". The sculpture is an abstract figure of a mother, father, and child holding a crescent moon, the symbol of Islam, around a sun, a symbol of the ancient Sumerian civilization of Iraq.

HIROSHIMA Japan

1945 The atomic bombs dropped on the Japanese cities of Hiroshima and Nagasaki changed the world for good. Hiroshima was slightly bigger than its atomic twin, Nagasaki, with a population of around 240,000. US commanders chose the city ahead of another shortlisted target, the historic city of Kyoto, because one of them had spent his honeymoon in Kyoto. Also, bomb experts felt that the hills surrounding Hiroshima might "focus" the blast and cause more damage. When "Little Boy" detonated, most of the city was destroyed – though some reinforced concrete buildings, such as the tower in the foreground, survived. Almost twice as many died as in Nagasaki – an estimated 140,000 in all. So total was the destruction that all communication with the outside world ceased, and it was 16 hours before the Japanese government learned what had happened – from the White House.

From the ashes Sixty years on, Hiroshima is a bustling, modern city. Its population has recovered to more than a million people, many of them manufacturing Mazda cars for export around the world. Its castle and shrine have been rebuilt and its annual flower festival resumed. Hiroshima has proclaimed itself a city of peace. And its Peace Memorial Park, commemorating the victims of the bomb, is now a popular and poignant visitor attraction.

DAL LAKE India

1986 This Himalayan idyll is the Dal Lake. Surrounded by the magnificent mountains of Kashmir on India's border with Pakistan, the area has in the past been sustained by a thriving tourist industry. Many visitors stayed in houseboats on the lake. Its shores were covered in gardens and orchards, the produce from which was sold in floating vegetable markets like this one on the lake itself.

Turning their backs
But the escalating conflict between India and Pakistan over the status of Kashmir has frightened away most of the tourists. Deprived of any incentive to retain the lake's beauty, local residents have begun building houses on its shore. The illegal construction has upset the hydrology, and effluent from the townships has turned parts of the lake into a quagmire of floating weeds. Having once been the hub of local life, the lake is increasingly an abandoned eyesore. Residents have turned their backs on its waters. The markets have closed – and even the old trading boats are sunk.

MEKONG RIVER Vietnam

Late 1960s When the Vietnam War began in the early
1960s, most of the region's rainforests were intact. Too intact for the
American military. The trees provided perfect cover for their enemies,
the Vietcong, who travelled the country bringing arms to rebels
and recruiting villagers. So the US took a leaf out of British counter-
insurgency methods against communists in the Malaysian jungles in
the 1950s. They decided to make the Vietcong visible with a virulent
weedkiller called "Agent Orange". Between 1962 and 1971 American
aircraft sprayed 50 million litres (11 million gallons) of Agent Orange
over a fifth of South Vietnam's forest.

Defoliation
Agent Orange's ability to defoliate large tracts of rainforest was soon clear across wide areas, as this photo of the Mekong River from 1970 shows. But about then doctors began reporting an epidemic of birth defects among Vietnamese villagers. It emerged that Agent Orange contained a contaminant called dioxin that caused foetal deformities and cancers. The spraying stopped. Today the forests are recovering – where they have not been cut down by loggers. But Vietnamese doctors report high rates of birth deformities even among the grandchildren of those exposed to Agent Orange.

MESOPOTAMIAN MARSHES Iraq

1973 The Mesopotamian Marshes in southern Iraq have for millennia been the largest expanse of wetland in the Middle East, covering an area of around 15,000sq km (5,800sq miles) in the flood season. They have sustained millions of birds and helped maintain fisheries in the Persian Gulf. They have also been home to the Madan people, whose ancient ways of hunting and fishing among the reed beds and waterways lasted for 5,000 years and were chronicled in Wilfred Thesiger's book *The Marsh Arabs*. Some say these marshes were the inspiration for the Garden of Eden myth.

The great drain Unfortunately, in the 1990s, following an old British colonial blueprint, Iraqi President Saddam Hussein constructed a huge system of dykes and canals to drain the marshes and divert the flows of the two rivers that fed them, the Tigris and Euphrates. The British aim had been to take water to irrigate fields in the desert; Saddam's was to flush out the Madan, who had rebelled against him at the end of the first Gulf War in 1991. Saddam's engineering project, inadvertently helped by dam-building upstream in Turkey, successfully drained most of the marshes before the decade was out, leaving behind massive salt pans and dead vegetation. The Madan fled to Iran.

Reversal After US and British forces removed Saddam, the Madan began to return, dynamiting the dykes and setting about the task of refilling the marshes. Within a few months large areas had reflooded and by late 2006 more than half of the former marshes had some water. But the water is much saltier than before, so freshwater ecosystems have not recovered and the Madan can no longer rely on the wetlands for drinking water.

BRANDENBURG GATE, BERLIN Germany

1988 Winston Churchill called it an Iron Curtain. The divide between East and West that opened up after the end of World War II was at first ideological, then economic, and, finally, a physical barrier separating the two halves of Europe. At its heart was the Berlin Wall, which ran through the divided city, and the most hated piece of that wall was the barricade constructed by the East German government in 1961 across the Brandenburg Gate, the only surviving ancient gateway into the German capital. The wall's purpose was to prevent the inhabitants of East Germany from fleeing to the freer and more prosperous West. Hundreds were shot in their efforts to climb the wall and make their escape.

Removing the barrier
Just as the barricaded Brandenburg Gate became a symbol of Soviet oppression, so the dismantling of the barricade became a symbol of the reunification of Europe. As the East German government fell in late 1989, one of its last acts was to allow its citizens to leave, which they celebrated by tearing down the wall, stone by stone, starting at the Gate. This picture was taken on 3 October 1990, when East and West Germany were formally reunited. Work to clean and refurbish the Gate began in 2000.

MOSTAR BRIDGE Bosnia

1955 This was once the largest single-arch stone bridge in the world. It was 30m (100ft) long, 20m (65ft) high, and made of 456 blocks of stone. When photographed in 1955, the bridge over the River Neretva at Mostar in Bosnia was already more than 250 years old. It had been built during the reign of the Ottoman Sultan Suleiman The Magnificent and joined the Croat and Serbian Muslim sides of the town. The bridge inspired many. Tourists came just to see it, and local men proved their bravery to their brides before marriage by jumping off it.

Destruction War came to Bosnia after the collapse of the state of Yugoslavia in 1991. As battles raged Croat tanks targeted the Mostar Bridge, and in November 1993, just a couple of months after this picture was taken, they destroyed it altogether. The loss of the bridge symbolically divided the town once again, although most Serbs had already fled. The ancient bridge was replaced by a rickety footbridge that served for more than a decade – a poignant metaphor for a fractured country.

Reconstruction After hostilities ended in 1997, Hungarian divers went in search of the old stones on the river bed and Turkish engineers began reconstruction of the bridge at an eventual cost of $13 million. The Mostar Bridge, complete with reinforced foundations, reopened in July 2004 amid great fanfare, as a symbol of a new beginning for the town and the country. But, sadly, few Serbs have so far returned to Mostar.

MOGADISHU Somalia

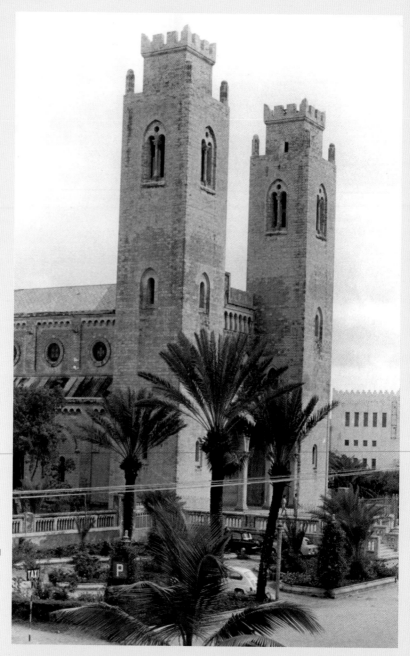

1966 Once, when Somalia was a functioning state and Mogadishu was its capital, the city had an active Roman Catholic cathedral on its main street. The cathedral, the largest in Africa, was built in Norman Gothic style by Italian architects in 1928, shortly after Mussolini's Italy took control of what was then called Italian Somaliland. It formed part of a grand plan for the capital. After years as a UN protectorate after World War II, Somalia gained independence in 1960, but retained a substantial Italian population, worshipping at its cathedral.

Failed state: fallen cathedral

Today independent Somalia is no longer a functioning state. Its government and infrastructure lie in ruins as warlords compete for Mogadishu, street by street. In 1989, armed insurgents killed the last Bishop of Mogadishu, Salvatore Colombo, while he was saying mass in the cathedral for his dwindling flock of parishioners. The cathedral itself was destroyed in later clashes during the country's protracted civil war, and many of its graves have been desecrated. Here, a couple of weeks before Christmas 2006, children play amid the ruins of the former cathedral of Mogadishu.

QALQILYA Palestine

2002 Qalqilya is a Palestinian market town on the West Bank. It is close to the pre-1967 border with Israel, and to the trans-Israel highway, which can be seen in the photo on the far left. Its inhabitants have lived for centuries by farming and, in recent decades, by providing labour for Israeli employers. The town also became a major centre for Palestinian refugees who lost their homes and land after the wars with the Israelis in 1948 and 1967. It has a population of some 40,000.

The wall In late 2002, after a Palestinian uprising that killed hundreds of Israelis, the Israeli authorities constructed a "security barrier" to separate Palestinian communities on the West Bank from their Israeli neighbours. Among the first places to be cordoned off was Qalqilya. In this satellite image, taken in June 2003, the barrier is seen as a white line virtually surrounding the town. It comprises an 8m- (26ft-) high concrete wall, a 4m- (13ft-) deep ditch, watchtowers, and an access road. Since its construction outsiders can reach Qalqilya only through an Israeli checkpoint. Farmers are cut off from their fields, workers cannot travel to work, and economic life in the town has come close to a standstill.

TWIN TOWERS, NEW YORK USA

Early 1990s Ever since their completion in 1972, the Twin Towers of the World Trade Center in New York were a symbol of successful global capitalism. Each was 110 storeys high, the tallest buildings in the world. Some 50,000 people worked in them every day, mostly for major financial companies such as Morgan Stanley. The centre had a troubled history: it was hit by fire in 1975 and a truck loaded with explosives was detonated in the underground car park of the North Tower in 1993. Islamic extremists later convicted of the bombing said their aim had been to destroy both towers.

Terrorism Despite this warning, nothing prepared New York – or the world – for the destruction of the Twin Towers on 11 September 2001. Terrorists from Al-Qaida flew packed passenger aircraft into each tower in turn, within 17 minutes of each other. Both towers collapsed in the hours that followed, with the loss of some 2,700 lives. The debris smouldered for 99 days. In early 2010 new buildings were beginning to emerge from the debris. The red steel girders are part of the new One World Trade Center (more properly the Freedom Tower) that will eventually reach 541m (1,776ft) and dominate the lower Manhattan skyline.

Computer waste Discarded TVs, computer monitors, and circuit boards are exported in their millions to developing countries to avoid strict rules intended to encourage safe recycling in Europe. Instead the "recycling" to recover valuable metals is done in backyard workshops, using dangerous low-tech methods. And the remains are dumped – as here in a lagoon in Accra, the capital of Ghana.

Leisure and Culture

What do we do when we are not making war? We make peace, we make art – or we go on holiday. Leisure reveals the human imagination untrammelled by practical needs. But it is big business, too. The annual holiday is a ritual for most of the world. And if you can, you have two or three or four. The skies and roads are full of people seeking inspiration and pleasure by going somewhere fresh to see or hear or experience something new.

Sometimes, admittedly, our demands are for basic sensory pleasures in a mildly exotic location. An area such as the Mediterranean – the world's top holiday destination. A firm favourite there is Benidorm, a Spanish fishing port whose white sandy beach has packed in so many holidaying Europeans in search of sun, sand, and sangria that it has become the "Manhattan of the Mediterranean". Or somewhere like the world's largest golf resort, created for Chinese businessmen on a construction site in Guandong.

More often than not we want spectacular scenery too. Cue the growth of Alpine resorts such as Zugspitze, once no more than a wind-blasted Bavarian meteorological observatory, but one that offered spectacular views across four countries. Then there is Zermatt, the exclusive car-less Swiss resort beneath the Matterhorn.

A sense of history goes down well too. "Heritage" as the tourist trade now calls it. Mount Rushmore attracts millions of Americans who queue to see the images of four famous presidents dynamited into the granite peaks of South Dakota. Bizarrely, so too does London Bridge, which was relocated from its River Thames home to a canal in Arizona and became one of the state's top tourist attractions.

But Londoners bemused by the attractions of a hand-me-down bridge have their own transported piece of history. Cleopatra's Needle, an Egyptian obelisk that spent the best part of 2,000 years beneath the sands of Alexandria in Egypt before being relocated to the banks of the Thames, is on most tourist routes through the city.

History, many would agree, is better when not transported. The Buddhas hewn from sandstone cliffs across Afghanistan qualified as true examples of cultural heritage. But that was before their desecration at the hands of the Taliban. Other genuine cultural treasures include the Mayan temples in the jungles of Central America and Luxor in Egypt. The latter, like Cleopatra's Needle, was rescued from the Saharan sands but, unlike the incongruous obelisk, has been left in its proper setting.

Context can be lost, of course. Often we end up despoiling what we come to see. The Great Wall of China is one of the great monuments of human history – famously visible from the moon. But the ancient barrier of the Middle Kingdom against Mongolian hordes is now corralled by the highways that transport the millions who come to see it.

Sometimes, however, tourist sites can even gain from their incongruous settings. The Eden Project, one of Britain's most popular attractions, is a celebration of the world's ecosystems and the wealth that mankind has created from them. The effect is enhanced by the fact that the site is an abandoned china-clay pit in Cornwall. And what price a sandy beach in the middle of Paris?

MOUNT RUSHMORE USA

1925 The Black Hills of South Dakota are imposing. The thrusting granite peaks are said to be America's oldest mountains, dating back at least 60 million years. The surrounding pine and spruce forests were once the domain of the Lacota Sioux people. But, like the state itself, they were little visited by outsiders until Doane Robinson, lawyer and state historian, had the idea of creating giant carvings of national heroes on the exposed rock-face of Mount Rushmore as a means of attracting tourists.

Reprofiling The first carving on Mount Rushmore began in 1927. But it took some 400 men, headed by the chief sculptor Gutzon Borglum, 14 years to complete the task. They removed half a million tonnes of rock, mostly by dynamite, to create the 20m- (65ft-) high faces of four of the US's most esteemed presidents: George Washington, who led the War of Independence against Britain; Thomas Jefferson, author of the Declaration of Independence; Theodore Roosevelt, revered early-20th-century president; and Abraham Lincoln, who campaigned to abolish slavery. Some three million visitors come each year as an act of reverence for their nation. The hard granite rock should ensure the survival of the faces: erosion rates are estimated at just 1cm (⅜in) every 400 years.

LUXOR TEMPLE Egypt

1857 For almost four millennia, two statues of Ramesses II, one of ancient Egypt's greatest kings, have stood at the gateway to the temple at Luxor in Egypt. The temple was the heart of the great city Thebes at the height of ancient Egyptian civilization. In this Victorian picture, the statues appear to be head-and-shoulders images. But look more closely. They are surrounded by piles of sand. The Sahara desert had been invading the temple for centuries, filling its rooms and piling up against its walls. There is more to these statues than meets the eye.

Clean sweep Thanks to assiduous restoration and the removal of several metres of sand from the temple floor, the statues are today revealed in their former glory. Far from being head-and-shoulder images, they are giant carvings of a seated Ramesses, depicting everything down to his toes. Luxor, on the east bank of the Nile in Egypt, is today one of the major tourist attractions of the world. It is visited by millions every year. Keeping Luxor free of invading sand is a constant job, as the sweeper at work in the foreground shows. But have restorers reached the true floor of the temple? Maybe not. The temple itself is built on a slight rise, and some believe that beneath it, still unexcavated in the sand, lie a previous temple and maybe more statues.

BAMIYAN VALLEY Afghanistan

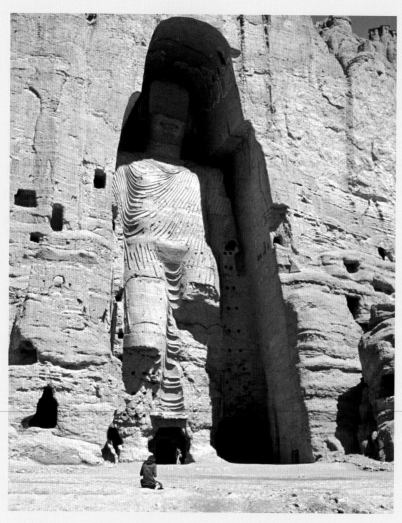

1987 Some 125km (80 miles) west of Kabul, in the heart of the Hindu Kush mountains, stood the world's tallest statue. At 53m (175ft) high, this representation of Buddha, hewn from the sandstone cliffs of the Bamiyan valley, counted as one of the world's greatest cultural treasures. UNESCO had designated it and a smaller companion statue as a World Heritage Site. The statues had been carved by the people of the ancient Buddhist kingdom of Kushan between the second and fifth centuries AD, when the caves in the cliffs housed a convent. The convent was abandoned after Islam came to the region. But though neglected and occasionally attacked thereafter, the statues survived for more than 1,000 years. Until the Taliban took power in Afghanistan.

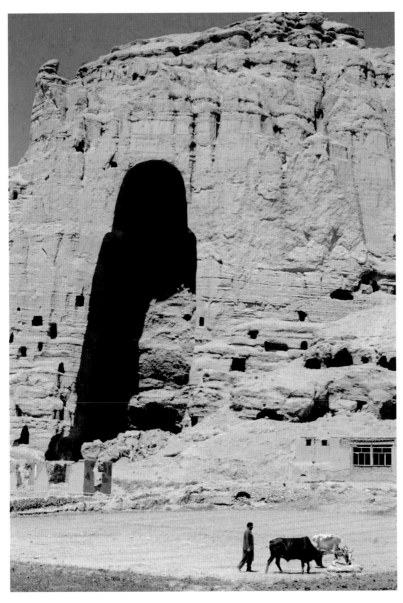

Desecrated idol

In February 2001, Mullah Mohammad Omar, the Supreme Commander of the Taliban, ordered the destruction of all statues in the country. Islamic teaching, he insisted, abhorred idols of all sorts, whether cultural treasures or not. Armed with dynamite and mortars, his troops spread out across the country to fulfil the order. And on 8 March the Bamiyan Buddha was reduced to rubble, leaving behind a bare gap in the cliff where the statue had stood for almost two millennia.

MAYAN TEMPLE Central America

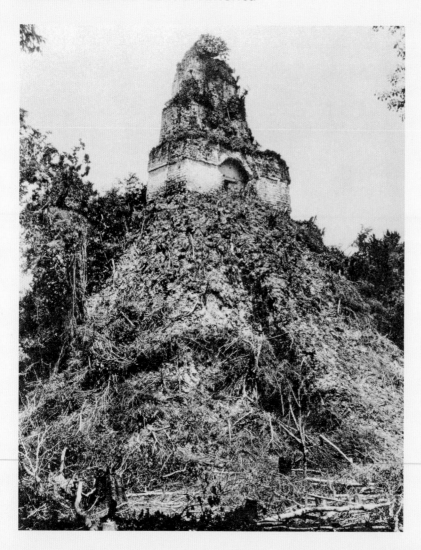

1955 Tikal is the largest of the "lost" cities of the Mayan people, who once flourished in the jungles of Central America. When first surveyed in the 1950s, many of the temples of Tikal were tall enough, at around 60m (200ft), to penetrate the tree canopy. But the huge pyramids on which they stood were entirely covered by vegetation.

Tikal was a major metropolis for almost 1,500 years, between the fourth century BC and the 10th century AD. The city spread across 60sq km (25sq miles). Besides temples and palaces, it built universities and giant irrigation schemes that watered farms cleared from the jungle. At its peak, Tikal had a population of up to 200,000.

Re-emergence After centuries forgotten in the Guatemala jungle, the main palaces and temples of central Tikal – like this one, known simply as Temple 1 – are now a major tourist attraction. But even today, after half a century of effort by archaeologists, only a fraction of the ancient city has been surveyed. Many lesser buildings lie shrouded in vegetation. One secret is becoming clear, however.

The Mayan civilization seems to have disappeared because of climate change. A long drought destroyed its farming systems. The people in the cities had no food – and eventually disappeared into the jungle to forage for survival. The "primitive" Native Americans discovered by Europeans centuries later were the survivors of a great civilization.

CLEOPATRA'S NEEDLE Egypt/England

c.1840 Cleopatra's Needle was erected 3,500 years ago in Heliopolis by an Egyptian pharaoh called Tuthmosis III. It was one of three, each 21m (69ft) high and made of solid granite hewn from a quarry near Aswan on the Nile. They were later moved to a temple in Alexandria that had been erected by Cleopatra – hence the name. But the obelisks then disappeared from history, buried in sand, for 2,000 years before archaeologists recovered them. One was given to Britain in 1819 to celebrate victories in the naval battles of the Nile and Alexandria. But the British government refused to fund its move to London, so it stayed for several decades behind a wooden fence on the outskirts of Alexandria, where this picture was taken.

Transportation In 1878 a public subscription finally raised £15,000 to pay to bring the obelisk to Britain. It set out in a specially made cigar-shaped iron pontoon, towed by a steam ship. In the Bay of Biscay the pontoon capsized in a storm and six crew drowned. But the partly swamped vessel was recovered five days later and eventually sailed up the River Thames more than three months after setting out. It was raised on the Thames Embankment, where it has sat ever since – a small piece of ancient Egypt on the streets of London. Two other "Cleopatra's Needles" also found their way to foreign parts in relatively recent times: one is in Paris and the other in New York.

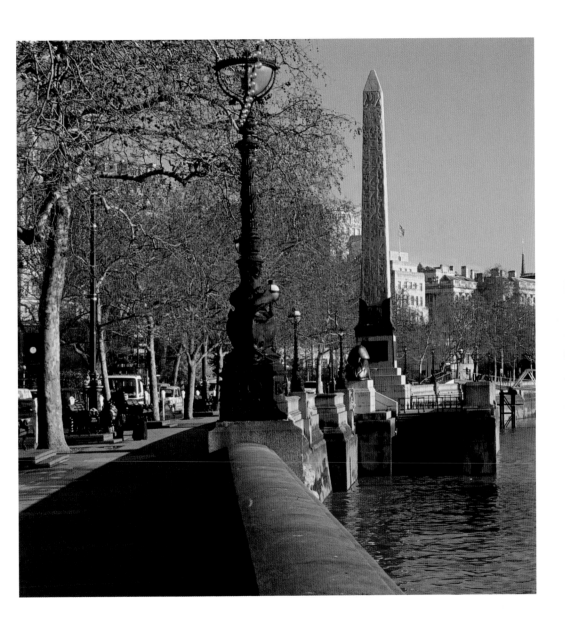

LONDON BRIDGE England/USA

1895 London Bridge spans the River Thames between the British capital's financial centre, the City of London, and Southwark on the south bank. The site has had a bridge since Roman times and until as late as 1750 it was the only bridge over the river in London. Old wooden bridges repeatedly burned down, but when stone versions were built in the Middle Ages they were colonized on top by shops and houses, and beneath by water wheels to grind corn. This version, made of Dartmoor granite, was completed in 1831 by the famous engineer John Rennie and was free of buildings. Even so, it began to sink into the Thames mud and was replaced in the 1960s.

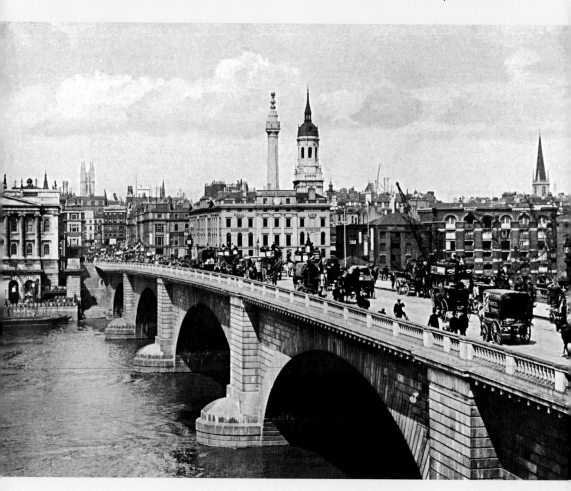

Transatlantic trip In 1968 Rennie's bridge was sold to US oil magnate Robert McCulloch for $2.4 million. He dismantled it stone by stone and shipped it across the Atlantic for reassembly (though not all of it: some stone was impounded in the UK in lieu of taxes). Some say that McCulloch believed he was buying the rather more distinctive Tower Bridge nearby. But it proved a smart purchase, anyhow. Today the old stone forms the cladding for a concrete bridge that spans a canal at Lake Havasu City in Arizona. It is the state's second-most popular tourist attraction, exceeded only by the Grand Canyon.

GREAT WALL China

1871 Once, the Great Wall of China was the farthest reach of the greatest empire on Earth: an impregnable barrier between the Middle Kingdom and the barbarian hordes in Mongola and beyond. The wall was constructed in stages from the fifth century BC. It is by some way mankind's longest structure, snaking along mountain ridges for some 6,400km (4,000 miles), and regularly dotted with forts and armed border posts. Even during the 20th century, many parts of the wall were in an empty landscape, such as this section in the Shuiguan valley not very far from Beijing.

Change and decay This is the same spot in the
Shuiguan valley today. The wall's location has been wrecked by
a major highway and tunnel. The Great Wall now appears almost
mundane in the face of modern civil engineering. A survey conducted
in 2003 found that a third of the Great Wall had disappeared, either
knocked down for roads and other developments, or scavenged for
stone. Another third was badly damaged. The wall is becoming a
victim of the greatest building boom ever seen on the planet. A fitting
riposte to the Great Wall, some might say. But will today's monuments
also be revered in two millennia?

QINGHAI–TIBET RAILWAY Tibet

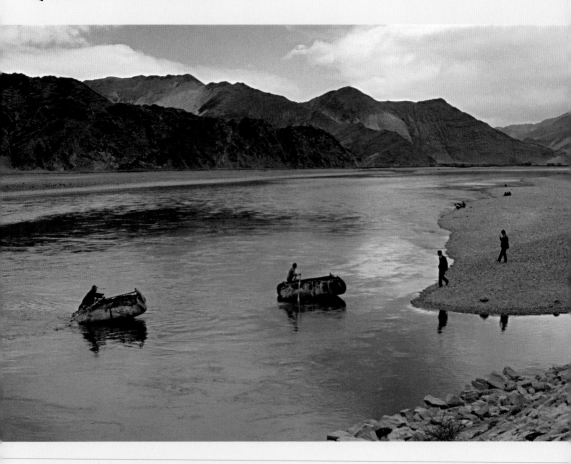

1987 In the mountains of Tibet, yaks are the most common animals. And yak hide is the material of choice for constructing small boats to cross the streams that drain from the glaciers of the great Tibetan plateau. Here, in the middle of the long dry season between monsoons, when the rivers dwindle, two Tibetans are using their yak-hide boats to cross the River Lhasa near the Tibetan capital of the same name. The image could have come from almost any era – certainly from a time when Tibet was an independent country. But since 1951 it has been under the control of China.

Crossing rivers And this is how the Chinese like to cross
rivers. In 2006, the new Qinghai–Tibet railway opened. It is the world's
highest railway; at one point it climbs to more than 5,000m (16,000ft)
above sea level. More than 1,100km (700 miles) long and built largely
across permafrost, it is designed to bind restive Tibet ever closer
to Beijing. Here it is crossing the River Lhasa over the new La Sa Te
Bridge, near its terminus in Lhasa.

ZERMATT Switzerland

1900 The Swiss village Zermatt sits at the head of a valley beneath the Matterhorn, the most famous mountain in the Alps. The village was once occupied exclusively by migrant pastoralists. But Zermatt's fate has been bound to that of the mountain since the peak was first climbed from there by a British team in 1865 – a conquest made all the more famous because four of the seven mountaineers fell to their deaths as they returned to Zermatt. This image was taken in about 1900, shortly after the completion of a railway up the valley. It shows early tourists visiting chalets in the village. The railway brought thousands of visitors every summer. In the days before skiing, most people just came to stand and stare.

Controlled growth

Zermatt is much bigger today, with a permanent population of more than 5,000, and more than 13,000 visitors' beds. The chalets have become small apartment blocks. But with no road access, it remains a village without cars. Electric buggies and horse-drawn sleighs are the only transport allowed, though the air is sometimes filled with the sound of an arrival at the heliport, where four helicopters will take well-heeled visitors for a tour around nine of the ten highest mountains in Europe.

ZUGSPITZE, BAVARIA Germany

1925 Zugspitze, in the Bavarian Alps right on the border with Austria, is the highest mountain in Germany. The mountain was not climbed until 1820, but in 1900 this observatory was built on its summit to provide weather reports for both Germany and Austria. At almost 3,000m (10,000ft), it was a remote and forbidding outpost that could only be reached by a two-day trek through the Partnachklamm Gorge.

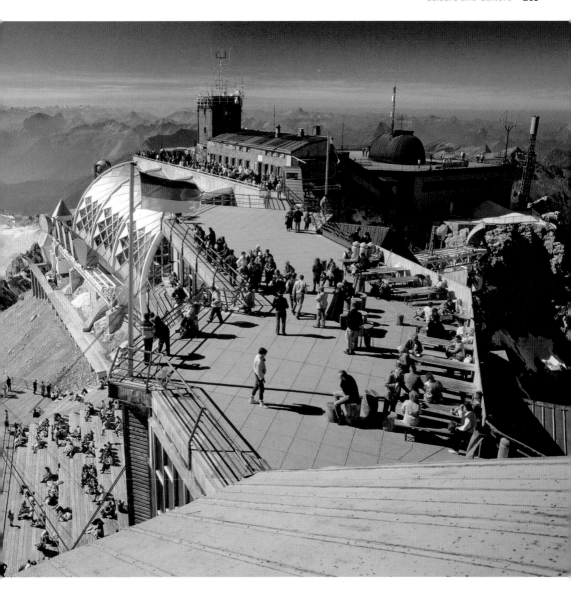

Access for all

The scene today is rather jollier. The wind-blasted outpost of meteorology has become a tourist attraction, providing stupendous views across four countries and some of the best skiing in the Alps. Access is easy these days. The summit of Zugspitze is reached by three different cable cars, two from Germany and one from Austria. And a cog railway improbably tunnels its way to within 400m (1,300ft) of the summit. Early risers get breakfast at the top, and the observatory even houses its own border checkpoint.

BRONX CITY HALL USA

2005 There are few bleaker places than the concrete roofs of high-rise public buildings, like this one above the ninth floor of the Bronx County Building in New York, home of the Bronx Supreme Court, and magistrates' and administrative offices. And when it rains, the water runs uselessly into the gutters, adding to local flood risks. Concrete and asphalt surfaces across major cities often trigger flash floods, because the water cannot soak into the ground. These surfaces also retain heat during the day, turning cities into "urban heat islands" that require extra air-conditioning of buildings in summer.

Green roof But there are solutions. This is the same
Bronx roof, planted up. The soil and plants capture rain to sustain
themselves and reduce runoff. The roof gardens help keep the
building and the wider city cool, and provide a pleasant refuge
for workers on breaks. Thousands of green roofs have been
created across the world's cities in recent years for their thermal,
hydrological, ecological, and psychological benefits. And the Bronx
greenery can even improve diet. The 15,000 plants up here include
lots of chives, ideal for spicing up the lunch sandwiches of the
workers toiling beneath.

CORNWALL England

1999 There can be few more unprepossessing sights than the sterile depths of a disused china-clay pit. The clay pits of Cornwall in southwest England have been in decline for many years, leaving behind huge, soil-less, waterlogged holes in the ground as well as vast piles of white spoil, and making parts of Cornwall into a virtual moonscape. But this pit, at Bodelva just north of St Austell, attracted the attention of Tim Smit, a former record producer turned aficionado of gardens. Improbably, he wanted to build in it the world's largest conservatory, to showcase plants and how we use them.

Eco-tourism
Thanks to a large national lottery grant and a lot of favours, this is the result – the Eden Project. The pit has become a vast terraced garden with, as its centrepiece, a series of geodesic domes. Inside, Smit has recreated entire ecosystems, such as rainforests and the Mediterranean, and displays teach visitors how the plants are used to provide food and clothing, and make fuel, building materials, and medicines. The largest conservatory, containing a huge array of giant tropical trees and plants that almost burst out of the dome, is 240m (790ft) long and 50m (165ft) high. Since it opened in 2001, more than 10 million visitors have descended into this Cornish Eden.

SHENZHEN China

2003 It's pretty much all bunker in this shot. José Maria Olazábal, one of the world's top golfers, takes a tee shot in the midst of a construction site somewhere between the two electronics boom towns of Shenzhen and Dongguan in southern China. Why?

Leisure landscaping

The first image was a publicity shot, of course. It was designed to show the transformation of the landscape during the eight months between that shot and this. The same golfer in the same spot now has no need of a tee to play this shot over the water to the green on the par-five 15th at the Missions Hills Golf Club. Olazabal designed this particular course, but it is one of 12 18-hole courses at what is claimed to be the world's largest golf resort, "only 20 minutes from Hong Kong". Golf is booming all across Asia, as the newly rich indulge in the sport of the Western well-to-do. And China, as in most things, is racing hard to catch up.

BEIJING China

2005 China is the premier construction site in the world. In recent years, the country has been consuming 70 percent of the world's concrete and 40 percent of the world's steel. Fittingly, the culmination of its building boom was the National Stadium for the 2008 Beijing Olympics, seen here part-built in 2005. The stadium, intended to house athletics and soccer events as well as the opening ceremony, eventually contained 100,000 tonnes of steel, making it the largest steel structure in the world.

A nest of steel

The stadium construction site grew into the "bird's nest" – the architectural image most associate with the Beijing Olympics. This night view of the giant steel nest – designed by a consortium of Swiss, Chinese, and British architects – was taken just before the Games began. The finished structure cost $420 million and contains 91,000 seats. Before the event, chief architect Li Xinggang promised that after the Games, it would "become the most important public space in Beijing". Things began slowly as paint peeled and the promised soccer club never moved in. But the stadium now draws many thousands of visitors daily, who pay $7 just to see the structure. And in early 2010 it was used as a snow "theme park" for skiing and tobogganing.

PARIS France

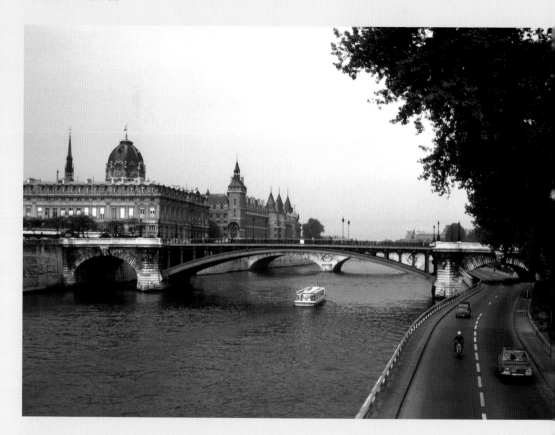

1975 Notre Dame Bridge over the River Seine is in the heart of Paris. Either side of the river, busy highways keep the city moving while tourist boats ply the river itself. Beyond the bridge, on the Ile de la Cité, is the giant Palais de Justice, the centre of the French legal system and a former palace.

Seasonal sands
But every summer since 2004, Paris has forsaken its riverside highways to turn the right river bank into the "Paris Plage". More than 2,000 tonnes of sand covers the tarmac; there are bandstands, and palm trees, and sunloungers, and marquees, and even a swimming pool. Millions of tourists are joined by Parisians who once deserted the city during August, but now stay to enjoy the fun.

JUMEIRAH BEACH, DUBAI United Arab Emirates

2003 What do oil sheiks do with their spare cash? The answer, increasingly, is that they put it into real estate. Not just buying old mansions and castles in Europe, but creating new real estate in the deserts, and even the seas, of the Middle East. Dubai, one of the United Arab Emirates in the Persian Gulf, is among the fastest-growing cities in the world. But besides turning into a great trading hub, the emirate is also becoming a holiday destination for the rich. In this satellite image, Dubai itself is toward the top right, centred on its creek. But stretching south along the coast is the Jumeirah Beach. And off the beach stretches the world's largest man-made island. Constructed in the shape of a palm tree, it is called The Palm.

Growing palms Three years on, the "palms" are multiplying. These complexes of private holiday islands are favoured by, among others, English football stars. The original complex, now known as Palm Jumeirah, has been joined by the larger Palm Jebel Ali to the south. And in this 2006 image, construction has just begun on the still-larger Palm Deira. If the original plans are completed it will eventually comprise some 8,000 houses on 250 small islands, and overall will be 14km (9 miles) long and 8km (5 miles) across. But since the credit crunch later phases have been put on hold.

AGADIR Morocco

1929 Agadir, for centuries a tiny fishing village on the Atlantic coast of Morocco, was reputedly the first sardine port in the world. Briefly a naval flashpoint for conflict between France and Germany in the run-up to the World War I, it relapsed again into anonymity when war was over. Here in 1929 it looked much as it must have done five hundred or a thousand years before. The main disturbance to life in Agadir over the centuries has been earthquakes. It has twice been destroyed by the shaking earth, in 1731 and again in 1960.

Winter sun
Agadir was entirely rebuilt after the 1960 earthquake. With Europeans starting to come to North Africa to enjoy the winter sun, the city fathers redesigned it on European lines with wide hotel-lined boulevards. Growth has been continuous ever since. Famed now more for its sandy beaches than its sardines, Agadir today has a population of 700,000 and is one of Morocco's top tourist destinations. The small fisher huts along the shore have been replaced by apartment blocks for European visitors, especially those seeking winter sun. An old American air force base, Al-Massira, has been converted to handle wide-bodied passenger jets and a new terminal accommodates three million passengers a year.

BENIDORM Spain

1963 Back in the 1960s, when most Europeans still holidayed in their home country, Benidorm was a small fishing village near Alicante, on the Mediterranean coast of Spain. Its domed church on a rocky peninsula, from where this picture was taken, provided good views of a particularly fine 6km (3.7-mile) stretch of white, sandy beach in three bays. But with travel companies set to start organizing cheap package holidays to the "Costa Blanca", its seclusion was about to end.

High-impact growth

Today in Europe only London and Milan have more high-rise blocks than Benidorm. The "Manhattan of the Mediterranean" has 300 buildings exceeding 35m (115ft) in height. It is said to have more hotel "stars" than the whole of Greece. Benidorm's permanent population of 65,000 grows tenfold during the holiday season when, despite the attractions of budget flights to an increasing number of cities across Europe, the Spanish coastal resort maintains its place among the Continent's most popular holiday destinations. More than a million British visitors alone come here to sample its casinos and discotheques each summer.

LAS VEGAS USA

2009 Now you see it... The desert city of Las Vegas is a byword for profligacy and rampant consumption, of fountains in the desert and lights blazing all night. The Las Vegas Strip is a byword for outdoor light shows whose glow is visible for tens of miles across the Nevada desert. Until recently, the lights had only been dimmed twice, when Dean Martin died in 1995 and when Frank Sinatra died in 1998.

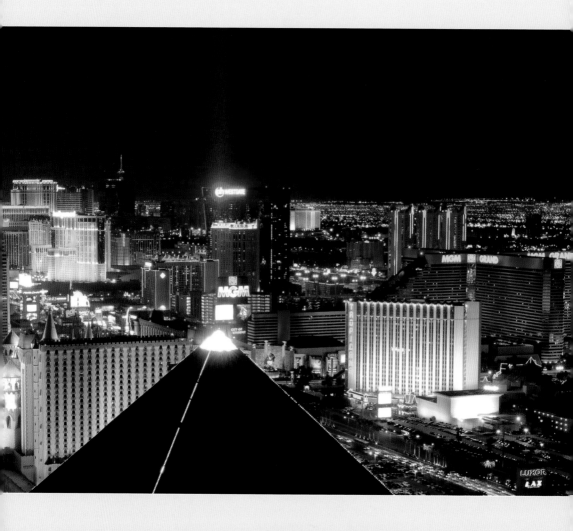

Lights out

But at 18.30 local time on 28 March 2009, the lights are switched off for Earth Hour. The annual event, organized by the environment group WWF, creates a global wave of darkness that sweeps across the world. It is intended to encourage businesses, communities, and individuals to take the simple steps needed to cut their energy consumption. Hundreds of millions of people round the world took part in 2009. Of course, in another hour, many of the lights are back on. Vegas will carry on as usual. But some energy had been saved, and some carbon dioxide emissions were prevented. Maybe most important, the hour was a symbol of what the world looked like before cities were flooded with light, and of how much of that power is wasted. Who needs those bright lights every night? Maybe they are a show that should only be put on for special occasions. Maybe most hours should be Earth Hours.

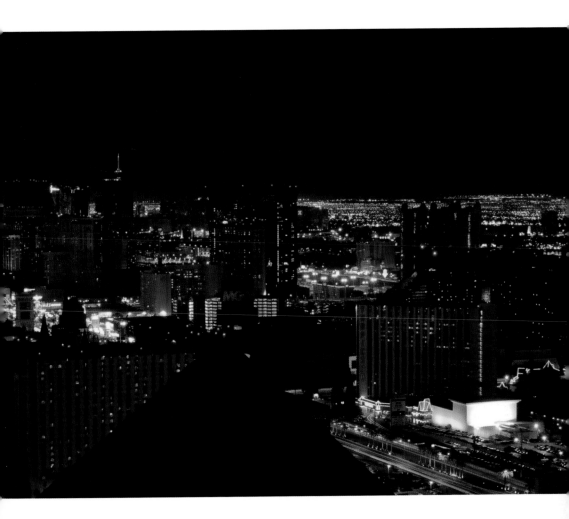

Index

Picture Credits